The Warehouse

The Warehouse

Workers and Robots at Amazon

Alessandro Delfanti

PLUTO  PRESS

First published 2021 by Pluto Press
New Wing, Somerset House, Strand, London WC2R 1LA

www.plutobooks.com

British Library Cataloguing in Publication Data
A catalogue record for this book is available from the British Library

ISBN   978 0 7453 4216 0   Hardback
ISBN   978 0 7453 4217 7   Paperback
ISBN   978 1786808 64 6   PDF
ISBN   978 1 786808 65 3   EPUB

Typeset by Stanford DTP Services, Northampton, England

Simultaneously printed in the United Kingdom and United States of America

Contents

Figures

A note on methods

This book is based on interviews conducted between 2017 and 2021 with Amazon warehouse workers and ex-workers at different levels (from seasonal associate to manager) and in roles covering most major processes and departments. It sides with workers struggling against the company, whether because they want to improve working conditions in its warehouses, or because they want to see Amazon gone, at least in its present form. My own politics and closeness to the labor movement, as well as the fact that Piacenza is my hometown, shaped the way in which I approached this book. Interviews reflect this positioning, although I also met several workers who were not directly involved in politics or who had a positive experience of working at Amazon. Most interviews were conducted in Italy, but I also spoke with people in Canada, the United States, Germany, and Spain. Workers at the warehouses of other e-commerce companies were also interviewed. To protect the identities of my informants, I used fictitious names, did not disclose the staffing agency they worked for, changed other recognizable details such as gender, age, or job when possible, and in some cases merged more than a worker into a single character in the book— or vice versa created two characters from a single interview.

I also downloaded and analyzed tens of thousands of comments left by Amazon associates on publicly accessible websites such as glass-door.com or Reddit, as well as YouTube videos and other content produced by warehouse workers. This material has been anonymized too. Workers were not the only sources. I conducted multiple site visits at warehouses and corporate fairs in three countries; attended both local and global trade union meetings, in some cases interviewing union organizers; spoke with members of worker-led collectives and alliances; analyzed publicly available corporate content such as training material, job ads, patents, letters to shareholders, and websites; finally, while I did not work at Amazon, I went through the selection process for a seasonal associate job and attended recruitment events in two countries. My research assistant Bronwyn Frey conducted ethnographic observations at re:MARS and other corporate events.

Acknowledgments

A number of students from the University of Toronto supported the research that went into the book: without Bronwyn Frey, Michelle Phan, Subhanya Sivajothy, Brendan Smith, Taylor Walker, and Adam Zendel this book would not exist. Erika Biddle aided with research, theory development, and in many other ways. My editor Matt Goerzen helped me develop and express my ideas: without him, this book would not even be in intelligible English. At times he worked in crunch sessions for which I am most grateful. Valentina Castellini sustained many such sessions, helping me shape, edit, and finalize the book.

The Bits, Bots, and Bytes reading group at McGill University, led by Gabriella Coleman, gave me priceless feedback on material that ended up in the book. The McLuhan Centre at the University of Toronto, led by Sarah Sharma, organized a workshop on an earlier version of the manuscript, which provided a number of great ideas and helped me fix many shortcomings. Friends, comrades, and colleagues volunteered their time to read parts of the manuscript or helped me improve it in a number of other ways, including Greg Albo, Hiba Ali, Nick Allen, Carina Bolaños Lewen, Tiziano Bonini, Olga Bountali, Antonio Casilli, Lisa Dorigatti, Nick Dyer-Witheford, Emine Elcioglu, Alessandro Gandini, Sam Gindin, Dan Guadagnolo, Omer Hacker, Into the Black Box collective, Tero Karppi, Anne Kaun, Tamara Kneese, Lilly Irani, Kira Lussier, Francesco Massimo, Rhonda McEwen, Massimo Mensi, Tanner Mirrlees, Fiorenzo Molinari, Andrea Muehlebach, Carlo Pallavicini, Julian Posada, Lilian Radovac, Nick Rudikoff, Liisa Schofield, Leslie Shade, Johan Söderberg, James Steinhoff, and Paola Tubaro.

The Institute of Communication, Culture, Information and Technology at the University of Toronto Mississauga provided the intellectual environment and material support that allowed me to write a book in the first place. Research was supported by an Insight Development Grant from the Social Sciences and Humanities Research Council of Canada, and a grant from UTM's Research and Scholarly Activity

Fund. A visiting position at the Department of Social and Political Sciences, University of Milan, allowed me to conduct part of my fieldwork. Last but not least, students in my seminars let me test some of my ideas on them, often helping me craft new ones—now you know why all those research-based assignments focused on Amazon.

My friend Barbara, who many years ago moved from our hometown Piacenza to Seattle and hosted me there many times, accidentally helped me see the two cities as connected to each other. My uncle Emilio assisted me especially when I needed to make sense of economic trends—of numbers, really. A serendipitous conversation with my friend Erica kickstarted the research that led to this book. Her lived knowledge as an e-commerce worker helped me connect many dots. I must also thank my old comrade Frenchi for noting that Piacenza "invented capitalism" and thus must be damned, and for mercifully providing the wine needed to process this information. David Shulman at Pluto supported my work since day one and helped me turn it into a book, including by patiently dealing with many delays.

I am humbled by the knowledge produced by thousands of Amazon workers across the globe, and can only hope that this book contributes to some extent to their struggle. Unions and worker collectives directly involved in the fight against Amazon gifted me with their precious time and ideas, including FILCAMS CGIL Piacenza, SI Cobas Piacenza, SI Cobas Pavia, FISASCAT CISL Piacenza, NIDIL Vercelli, FILT CGIL Roma e Lazio, Transnational Social Strike, Amazon Alliance and UNI Global Union, Warehouse Workers for Justice, Warehouse Workers Centre, and Amazonians United. Both Amazon and Zalando's press offices helped me visit their fulfillment centers.

Most importantly, I am enormously grateful to all the workers and ex-workers, at Amazon and beyond, who shared their experiences and ideas with me, even when that put them in uncomfortable positions. Some became friends. Many will disagree with my ideas. All should get a free copy of the book: if for some reason you do not receive it from me, I hope you will manage to ~~steal~~ read one in the FC.

Bobbio, summer 2021

1

Relentless

It takes just 15 minutes to drive from my hometown of Piacenza to the oldest and biggest Amazon warehouse in Italy. Taking the A21 highway westbound, the warehouse appears on the right just before the exit for the small town of Castel San Giovanni. Codenamed MXP5, the massive building is low in height but spans nearly 400 meters. Rectangles in different shades of gray decorate the exterior, capped by an orange line near the top—the same orange used for the smiling arrow that underlines the massive Amazon logo identifying the warehouse to passing motorists. Only a parking lot for the workers' cars and a designated yard for the continuous flux of trucks separate the complex from the busy highway. For years I used to drive by this place, back and forth, every day, to my workplace down the road in Stradella, before I moved to a new job and a new country. But back then the warehouse was not there. It appeared during an explosion of growth in the early 2010s, when an entire stretch of countryside in the Po Valley was reinvented as a sprawling logistics hub—strategically positioned to serve major markets like Milan and Turin. Hundreds of hectares of prime farmland are now covered by the warehouses of IKEA, H&M, FedEx, Zalando, and probably every other major global distributor you might think to name. Amazon's facility opened in 2011, and new companies edge in every year, bringing with them more concrete, more roads, more trucks, more workers, and more spotlight for the once forgettable Piacenza.

If I pulled over and got on my phone, it would take me mere seconds to open Amazon.it and order something. The place spits out hundreds of thousands of orders per day, probably moving up to a million items when running at full steam. For years its speed set the bar for Amazon warehouses all over Europe. This means a total of more than 3,000 workers organized in shifts, 24/7, under this roof. My order would not go straight to them, though. It would move at the speed of light from my phone to one of Amazon's data centers, maybe the one in Ireland, where it would be analyzed by the company's algorithms in its cloud computing servers. If I were ordering something to my childhood home and the product, say a new notebook, was stored and available in MXP5, the algorithm would ask the workers inside the warehouse to retrieve, pack, and ship it. If I were a Prime subscriber, the item would be in my mailbox tomorrow. Most people encounter Amazon exactly in that way: as consumers. The very name it gives to its warehouses, "fulfillment center" or "FC," signals that the goal of the company is to fulfill people's desires, or to create new desires that can be fulfilled by e-commerce. Thanks to its ability to take care of everything, from sourcing products to last-mile delivery, Amazon is becoming synonymous with the market, pretty much in the same way in which Facebook tried to become the internet. Buy more stuff, faster, more conveniently, cheaper, no need to look elsewhere.

According to theories of consumption, it is desire that motivates us to purchase commodities. This desire can be thought of as either an artificial thing forced upon us by marketing forces, or as reflecting a need to define ourselves through the things we buy.[1] Not to mention much more mundane needs: as demonstrated during the coronavirus pandemic, the consumption of food, clothing, or pharmaceuticals cannot be taken for granted. Our ability to consume is contingent and depends on complex global supply chains that can break down. Wherever our desire for consumption comes from, it must be fulfilled. But who we are as consumers cannot be separated from the society we are entangled in, a society where Amazon works to position itself as a global fulfillment monopoly, to overcome the distance between its 300 million customers and the commodities they need and desire. Most of these customers are introduced to Amazon as a safe, convenient, and, in some areas, even necessary way to access the commodities they desire—the mainstay of today's one-click instant consumerism.

But this is not how I was introduced to the company. Before I ever ordered anything from Amazon, I had read countless articles in *Libertà*, the local Piacenza newspaper, heard untold stories from people who work there—including old classmates and friends—and discussed it ad nauseum at dinner parties and political meetings alike. Amazon is a constant presence in the territories surrounding its warehouses, from billboards advertising job openings and signs directing truck drivers, to local news articles featuring both mayors who praise the creation of jobs and environmental groups who denounce the polluting effects of increased traffic. In Piacenza, Amazon became a heavy presence before the area was even an eligible destination for delivery. Boxes from MXP5 went out to the more modern and busier metropolis of Milan, while we got the jobs, the downward wages and working conditions, and the environmental degradation. Eventually, we got some national—and even international—attention as well. Who would have thought that a strike at a warehouse in Castel San Giovanni, or anything else happening near Piacenza for that matter, would make it into *The Washington Post*, the American mainstream newspaper owned by Jeff Bezos himself: Amazon's founder, former CEO, and biggest shareholder?

But that's just what happened. On November 24, 2017, hundreds of MXP5 workers went on strike. Unions had entered the company just a few months earlier. From the perspective of those in Piacenza, this looked like an unremarkable affair—yet another strike at a local logistics facility, an industry whose workers are in a perennial state of ebullition that at times erupts into open revolt. But from a zoomed-out perspective, this was more novel: MXP5 employees were among the first in the world to attack Amazon's empire head-on. In the intervening years, organizing and strikes at Amazon have grown, both in Europe and North America, making it a hotbed of struggles—a symbol of both capital's unchecked power and worker resistance. The first MXP5 strike coincided with Black Friday, the day when many retail stores offer sales and discounts, and a big day for Amazon in several countries, Italy included. On the very same day, Bezos' fortune attained a whopping $100 billion (US), making him the wealthiest person on Earth, at least at the time. Plenty of reasons for Piacenza to achieve momentary global fame.

UNBOXING AMAZON

In the opening lines of *Capital*, Karl Marx famously wrote that: "The wealth of societies in which the capitalist mode of production prevails appears as 'an immense collection of commodities.'"[2] Yet, of course, a collection of commodities is nothing without their movement from production to the market. With the warehouse, it is the immense *circu-*

lation of commodities that produces wealth. After all, if the warehouse is where commodities are accumulated, they must be moved around if they are to fulfill consumers' desire. They must be kept in flow, restless. If the stuff were stuck inside the warehouse, this would spell death for commodities and their value.

But commodities do not circulate by themselves. What lies in between your home delivery and Bezos' bottom line is a series of technological systems that organize the company's massive workforce, speed up work, and contribute to making jobs more precarious and unstable. It is human labor that keeps the commodities moving, in concert with and under the direction of a complex infrastructure of both software and machinery. When we receive a box from Amazon, we do not always think of all the planes and trucks, all the data centers, all the human labor that went into delivering it. We do not always think of workers like Giulia. When I first spoke with her, Giulia had recently lost her job at MXP5 after a few months as a seasonal worker. Hired through a staffing agency, her contract ended after the winter delivery peak and was not renewed. As we met in a coffee shop not far from the warehouse, she opened our conversation by explaining the mismatch she felt as both an Amazon customer and a former FC worker. Like many others, she had walked in both shoes:

> If you think about it, those who [order from Amazon] don't know who's behind it, behind that package you receive at home. The only time I ordered something from Amazon, I received one of the boxes I used to pack. I hung it to the wall and added a caption: "lest I forget."

Giulia told me a story that I could have heard from any of the dozens of Amazon workers I've spoken with. Amazon's e-commerce operations rely on around 200 fulfillment centers globally. Each spans hundreds of thousands of square meters and employs several thousand workers known in corporate lingo as "associates," and informally as "Amazonians." These "modest-looking buildings," as an Amazon commercial describes the FCs, are often highly visible when you drive past them. But what happens inside or around them is often less clear. The warehouse's walls are not transparent, and capital always does its best to make workers, the human side of its operations, invisible. The fact of automation, for instance, allows corporations to present what they do as the near-magical work of machines. But the removal of the labor of humans from the consumer's view has not made it disappear. In a book on the dark side of the digital revolution, Italian philosopher Roberto Ciccarelli reminded us that labor is "the faculty that feeds circuits and automatisms [...], the capacity that allows the production of a commodity and its value." In other words, it is labor alone that allows a firm like Amazon to exist.[3] And unlike commodities and machines, labor is something that no capitalist can fully own. It is workers' labor power that can be purchased, controlled, and disciplined, but its full potential can only be possessed by workers. Amazon does its best to purchase labor power. With 1.2 million employees as of early 2021, it is one of the biggest private corporations in the world, trailing only behind Walmart with its over 2 million workers. This number is even more astounding when you learn that just ten years earlier, in 2011, Amazon had only 30,000 workers. These workers are located across the globe. A home base in Seattle forms the center of a massive global network of offices, campuses, data centers and warehouses—like MXP5—that span North America, Europe, and Asia.

Making sense of the company is a titanic task, and not simply because of its size. Bezos often says that Amazon has a "willingness to be misunderstood." That is to say, the company does not care if competitors, investors—or the general public—are unable to comprehend its strategies.[4] Amazon directs this same party line at workers, unions, and public institutions alike. For instance, in an unprecedented move after the 2017 strike, MXP5's management no-showed a meeting called by a representative of the national government to discuss a possible agreement between unions and the company. They did not offer a rationale. "We do not expect to be understood," is what they told union representatives as justification for shunning the meeting. Perhaps, then, we must go to the workers themselves and rely on their knowledge to help us unbox the reality behind the smiling arrow logo.

MXP5 associates represent but a small chunk of the company's global workforce, which is dispersed throughout its network of warehouses and other distribution centers. When they meet fellow Amazonians from other FCs, it is mostly at national or international union meetings, or in the online forums where thousands of workers gather to discuss their experiences, share tips for survival, and, often,

vent against the company. Nevertheless, their experiences are gener-
alizable in a way that might not be true of other companies. Amazon
is highly centralized in the design and operations of its warehouses,
and thus the human costs accrued by the company's accumulation of
wealth and power speak to a global system of exploitation—a system
that one can find in many different local contexts, from a suburb in
America to a de-industrialized European metropolis. Local differences
matter, but workers in Castel San Giovanni encounter the same technology, workplace culture, and political strategies faced by Amazon
employees in other countries and continents. They struggle against
the same power structure and organization of labor. Everywhere, FC
work is based on the use of Amazon's innovative technology and, at
the same time, on archaic forms of despotism reminiscent of the factories of yore. The flashy corporate image familiar to customers is in fact
only one side of the coin. Amazon incarnates the disruptive power of
technology and the excesses of modern consumerism, and at the same
time beckons a new degradation of work. Amazon capitalism, as the
economic system underpinning the company has been called, is rapa-
ciously predatory of workers, other businesses, and the environment.[5]
It can also be resisted.

AMAZON DOESN'T STOP

Amazon prides itself on being relentless. This is a recurring word in
Amazon's corporate history and it appears over and over again in Bezos'
speeches and annual letters to investors. In fact, Bezos once envisioned
naming the company Relentless, and if you type in "relentless.com"
in your browser, it will redirect you to amazon.com: the company
still owns the domain. Amazon maintains that it remains relentless
because, in Bezos' words, "it's still day 1." That is, one of the biggest
corporations in the world still thinks of itself as a start-up company
that needs to move fast and never settle for the status quo. And indeed
the 24/7, always-on, crunch-time obsessed, burnout-prone culture that
characterizes tech start-ups is rampant throughout the company, and
is even impressed upon the company's warehouse workers. The technologies, management techniques, and cultural elements that impose
productivity onto start-up coders in Silicon Valley campuses are
increasingly applied to more and more sectors of the workforce. Bezos

has repeatedly emphasized his focus on making Amazon a lasting influence in the evolution of consumption, but he rarely describes how his company is also influencing the broad practice of work at a deep register. Amazon has the ability to reshape the fabric of our societies, redefine the role of corporate power, and shape the future of work to the advantage of capital.

The company has expanded dramatically from its rather unassuming origins. Bezos founded it in 1994 as an online bookstore, planning to take advantage of the new opportunities opened up by the commercial internet. His e-commerce ambitions were broader in scope, but books were a perfect test: brick-and-mortar bookstores are constrained by space and can carry thousands of titles, at best—a fraction of what you can squeeze into a warehouse. Books are also homogeneous in shape and weight and thus easy to store, and come with a well-established coding system. Initially, the company operated out of a single warehouse near Seattle. But the idea had always been much bigger. Bezos planned to digitize a business model that had been around for a bit: mail-order catalogs, a product of 19th-century modernity that created massive companies such as Sears in the US, Postalmarket in Italy, or Eaton Co. in Canada. By 1998, Amazon was already selling music and DVDs, to which it soon added home goods, toys, and video games. In the early 2000s it expanded to health and personal care products, gourmet foods, and sporting and outdoor goods. In 2005 it launched Amazon Prime, a subscription program that gives access to special services such as free one-day delivery—today, it is available in hundreds of cities worldwide. Amazon now sells everything, from cookies to electronics, bicycles and home appliances. In the US, it has grown to occupy about half of the entire e-commerce market, according to recent estimates—the quota is even higher in countries that lack competition. In Italy, for example, Amazon's market share reaches 60%.

But Amazon has become much more than just an e-commerce company. And e-commerce is not where the company's profits come from. This is an important, if difficult, reality to grasp. Indeed, Amazon's size and internal differentiation make it difficult to capture it in its entirety. Amazon Web Services (AWS) is the biggest provider of web space and computational power in the world. Services like Netflix, Pinterest, Airbnb, and Uber are run on these servers. Amazon also develops a number of commercial software technologies, such as its

"1-Click" online payment service. The service Amazon Mechanical Turk (AMT) allows businesses to hire remotely located "crowdworkers" to perform tasks that computers can't perform without human intervention, such as identifying content in a picture, tagging and cleaning data, or writing product descriptions. The platform breaks down these tasks and outsources them to a distributed workforce— anyone with a computer connected to the internet can sign up and work for AMT.[6] With its Rekognition program, Amazon sells surveillance technology to governments. It produces digital gadgets such as the Kindle e-reader and the Fire tablet. Its Echo smart-home device runs Alexa, a virtual assistant underpinned by natural language processing algorithms. Amazon also owns and operates a streaming platform, Prime Video, and has become a major producer of films and TV series through its Amazon Studios. It owns a chain of automated convenience stores called Amazon Go. Its many more subsidiaries range from game streaming platform Twitch to organic supermarket chain Whole Foods.

Amazon Web Services is the real money maker. Each dollar Amazon spends operating AWS generates ten times more profit than a dollar spent on its other ventures. That is to say, while the FC and services like Prime turnover more money than AWS, those services are not always profitable. AWS, on the other hand, generates a massive stream of money, enabling the company to expand its flagship e-commerce operation by relentlessly building new warehouses across the globe, steadily securing near-monopolistic positions in more and more national markets. Only Chinese e-commerce behemoths Alibaba and Tencent approach Amazon's size.

Amazon applies its concentrated economic power toward technological change too. This simple fact means that it has the ability to deeply influence the way in which we work. Not only in its warehouses, but throughout our societies. The reason is simple: from the widespread robotization of its FCs to its deployment of algorithms to monitor workers and extract valuable data from their labor, Amazon is relentless in increasing the rate of technological innovation in its warehouses. And this means other companies adopt similar technologies in their attempt to catch up with Amazon and uproot the company from its dominant position in the market. New technologies, more sophisticated, more pervasive. More workers and more precarity, too.

Relentless, also, is the circulation of billions of commodities, which move across the globe to converge on the warehouse, where they rest for a few hours or weeks on its shelves, and then depart again to reach new destinations. Their movement must be seamless, fast, efficient. Frictionless, as economists would put it. Workers are the most problematic factor in this equation, and thus must be carefully controlled and governed lest they generate friction, slowing down or even stopping the movement of commodities.[7] Amazon is at the forefront of digital capitalism, which means it excels in the corporate use of digital technology to maximize the private accumulation of power and money. But because workers play such a crucial role, Amazon is also at the forefront of a global offensive against labor. This offensive is fueled by the technology Amazon designs and deploys in the warehouse: Amazon has made precise choices to use technology in service of its obsession for control, for speed, and of course for money. Its technological infrastructure is aimed at workers' exploitation rather than emancipation.

The brutal reality of work at Amazon is no mystery. Even Amnesty International has issued a report about the poor conditions encountered by workers in Amazon warehouses, concluding that it is time for the company to protect the rights of its workers—for example, by respecting their right to unionize.[8] The undercover investigation of an Amazon fulfillment center has practically become a full-fledged genre for journalists in both Europe and North America. And Amazon serves as shorthand for the awfulness of contemporary work in a rich tapestry of internet memes. In Italy, *Lercio* (an Italian sarcastic news organization roughly equivalent to *The Beaverton* in Canada or *The Onion* in the US) once titled an article: "Amazon employee boxes and ships herself home to get a few minutes of break." These jokes do not come out of nowhere. Bezos himself has acknowledged repeatedly that "it's not easy to work here,"[9] emphasizing that sacrifice is a part of what's requested of Amazon employees, from the warehouse all the way up to the shiny executive offices in Seattle.

Many books have been written about Amazon. Often they are those ubiquitous business manuals found in airport bookstores, targeted at corporate executives—or wannabe corporate executives (one provides a blank page at the end of each chapter for "Reflections and ideas to consider for your company."[10]) The authors, who are typically business journalists, consultants, or business school professors, tend to present Bezos as a contemporary hero of entrepreneurship and innovation—but even they cannot avoid mentioning his temper and disdain for workers. The white-collar workforce is subject to his ire, too, we are told. Business journalist Brad Stone, the author of one of the best of these books, reveals that Bezos is well known for screaming at employees in public or firing them in the elevator. Employees interviewed by Stone described the company's former CEO as cruel and lacking empathy, as someone who treats workers as expendable resources.[11] A 2015 *New York Times* article also unveiled a cut-throat and punishing workplace culture in Amazon's offices, a place where employees routinely cry at their desks and where metrics are used to evaluate and pit employees against each other. Amazonians told the journalist they felt like they had to leave the company because of health crises and pregnancies.[12] These stories, mind you, were about engineers and executives.

Now imagine warehouse workers. They work under physically punishing rhythms, dictated by distant corporate algorithms which organize their labor. A pervasive surveillance system monitors their productivity at every step. The valuable information generated by their labor is captured and monopolized by Amazon's software systems, and then fed to the machines that run the warehouse and organize fulfillment processes. Employee turnover is high, by design, as the warehouse quickly discards and replaces workers worn out by the dictated pace. Precarity is promoted and exploited by management, enabling it to adjust the size of the workforce to meet the always fluctuating demand of the market. Amazon's management techniques take explicit and subtle forms. Supervisors interface with warehouse workers both directly and with the aid of digital technology. Meanwhile, the company cultivates a workplace culture aimed at convincing workers that warehouse work is special and fun. A combination that can be unsettling. And this is just the warm-up: Amazon has plans for a future FC that is even more technology-intensive, where its domination over the workforce is even stricter, and the labor process is increasingly automated.

The coronavirus crisis has enhanced the visibility of these dynamics, as Amazon increased its workforce and e-commerce quickly grew to become one of the dominant areas of the retail industry. In a sense this is nothing new, as corporations often exploit and benefit from systemic crises—Naomi Klein called it "disaster capitalism."[13] But few have been as successful at this as Amazon. Indeed, Amazon owes much of its success to maneuvers during three global crises. First, it survived the burst of the dot-com bubble in 2000, which put many competitors out of business and allowed the company to move out of its niche market position and into the center of the digital economy. The bubble had been created by massive influxes of venture capital into companies that focused their operations on the internet. Many were early e-commerce companies, such as the pet supplies retailer Pets.com. While others went belly up as the bubble burst, Amazon flourished, having managed to secure a sufficient pool of capital to allow it to float above the waves of bankruptcy and insolvency. In the fourth quarter of 2001, when hundreds of internet-based companies were folding, it turned its first profit, paying its owners one cent per share. Second, in the aftermath of the great recession of 2008, Amazon grew its workforce by tapping into a newly formed mass of workers made precarious by unemployment, debt, and the crumbling of labor rights that followed the crisis in Europe and North America. This new reserve army allowed the company to increase its global workforce from 20,000 in 2008 to over 100,000 in 2013, with a growth rate above 30% per year and up to 66% in 2011.

Finally, the coronavirus pandemic of 2020–2021 served as the ultimate perfect crisis for Amazon, as it had the dual effect of both multiplying the market for e-commerce and web services, and throwing millions of workers into unemployment. Amazon hired hundreds of thousands of new warehouse workers to cope with the bump in sales caused by the pandemic. In early 2021, it had grown its global workforce by 62% from 2019, and had increased its revenues from $280 in 2019 to $380 billion in 2020. Bezos' personal wealth when he resigned from his CEO position in early 2021 approached $200 billion: one of the vastest fortunes ever accumulated by a single human being in the history of human civilization. Amazon was perfectly positioned for the new reality brought about by the virus. The sharp increase in internet usage by those confined to their homes meant Amazon profited any time demand increased for a company that relied on its AWS servers. For instance, the teleconferencing company Zoom—an AWS client— multiplied its average daily users with offices and schools moving to online meetings and lectures. Most importantly, widespread lock-downs and fears of contagion boosted online shopping orders across the globe, especially in countries where Amazon was already the dominant player in the e-commerce market.

Overnight, Amazon workers became essential. In March 2020, at the height of the first wave of the pandemic in COVID-19-ridden Northern Italy, a courier shot a cell phone video while delivering for Amazon. It quickly went viral over messaging app WhatsApp: "Amazon doesn't stop," the worker can be heard saying through his medical-grade face mask, "don't worry, you will receive your damn Hello Kitty phone cover on time. Fuck you!"

Worker rage against Amazon consumers is understandable in this case. But what about rage against the company itself? Amazon's revenue increased so vastly over the course of 2020 that at the end of the year Bezos could have personally given every single Amazon employee $100,000 and still maintained his personal wealth at its pre-COVID-19 level.[14] The pandemic hit MXP5 hard, generating more trouble, more sickness, more fear for the workers. Piacenza was one of the early epicenters of the global pandemic, with 1,000 deaths in two months in a small province with a population under 300,000, and the warehouse soon recorded cases of virus infection. Nevertheless, during the first lockdown, when the sound of ambulances roaming deserted roads was many citizens' primary contact with the outside world, MXP5 was working 24/7 to keep up with the increased demand. But Amazon's record year did not translate into major improvements for the workers. Thanks to union presence and to the 2017 strike, MXP5 hirees remain among the small slice of fortunate Amazon FC workers who have managed to negotiate improvements, for instance, obtaining a substantial pay bump for night shifts. During the pandemic, like other Amazon workers across the globe, they received a minor salary increase (or "pandemic pay") for a few months and a small cash bonus. But in spring 2020, MXP5 employees had to mobilize again, this time with a long strike, to win basic things like the provision of adequate personal protection equipment.

In the meantime, Amazon was just making more money. In the video game *You are Jeff Bezos*, players are tasked with spending his wealth, which is no easy task. Just 10% of Bezos' wealth allows you to double every Amazon employee's salary. What about the remaining

90%? For instance, you can end homelessness in the United States (for a mere $20 billion), or even pay your personal taxes (a hefty $57.72 billion).[15] The company's accumulation of capital is even more incredible. Valued at $1.65 trillion dollars, Amazon is worth more than the GDP of rich countries like Australia or Canada, and not far off from Italy's $2 trillion. Barring antitrust legislation, the company is projected to grow even bigger. This beast, to quote Dante's *Inferno*, "can never sate her greedy will; when she has fed, she's hungrier than ever."[16]

AN OUTPOST OF DIGITAL CAPITALISM

Piacenza does not appear in this story only because it's my hometown. The city has found itself simultaneously at the center and at the margins of global networks of the trade before. In the late 16th and early 17th century, Piacenza was chosen by the all-powerful Genoese bankers to host their quarterly fairs. Soon enough, "the relentless heart [...] of the entire Western economy beat here at Piacenza," as put by French historian of modernity Fernand Braudel.[17] These fairs were not tumultuous street festivals, but rather meetings where a few businessmen from all of Europe exchanged what we would now call financial instruments: letters of credit, debts, and remittances. This was a major episode in the fortunes of the Genoese bankers, who by then were funding the Habsburg after the royal house had defaulted, bankrupting the Fuggers, the German banking family, in the process. It was also a major episode in the history of capitalism: economist Giovanni Arrighi called the rise of finance under the Genoese bankers "the first systemic cycle of accumulation."[18] Piacenza was chosen as the site of these fairs due to its convenient position at the crossroads between the Po river and the Via Aemilia. And probably also, added Braudel, for its "discretion."

Amazon must have had similar reasons in mind. Once again bustling and yet discreet and removed from the public eye, central and yet peripheral, today's Piacenza moves commodities rather than letters of credits. Finance is far away—in Milan, London, and New York. But Piacenza is still at the center of a network of highways and railroads that extends to urban centers in Northern Italy and beyond. It is now a major logistics hub right in the middle of the Po Valley, the relentless heart of a commodity flow that connects to other centers of the global economy.[19] Piacenza plays small cousin to other areas with a heavy Amazon presence, like the Inland Empire near Los Angeles, the Peel Region in the Greater Toronto Area, or El Prat de Llobregat near Barcelona. These metropolitan hinterlands, to use Phil Neel's term for the booming de-industrialized periphery, are spaces punctuated by logistics complexes, factories, sprawling suburban residential areas, residual rural areas and highways.[20] The warehouses, the relentless hearts of Amazon's e-commerce empire, are here, far from the downtown high-rises where commodities are designed or marketed.

MXP5 and many other warehouses in Amazon's network benefit from their proximity to wealthy urban markets. But their location is also meant to strategically exploit cheap labor forces, sometimes with startling implications. For instance, workers in Polish FCs may make as little as 3 euros per hour, and yet the packages they ship serve a German market where the same workers would make 11 euros per hour.[21] In other cases, FCs are simply placed in areas where prospective workers are abundant. Many of the warehouses that service the Greater Toronto Area are located in Brampton, a rapidly growing town that is home to a major South Asian community. Thus it's these racialized folks who must bear the brunt of the environmental impact that comes with heavy traffic to and from warehouses, as well as the human impact of the COVID-19 outbreaks that have affected Amazon FCs in the area—crowded workplaces with limited access to sick pay. The company exploits and perpetuates the injustices that come with the geography of our cities and countries.

From the perspective of those working in MXP5 and other FCs, the company's Seattle headquarters feels distant and out of reach. So do the national offices in Milan, which veteran associate Peppino described to me as "an inexpugnable building full of security cameras and bodyguards," where suits who barely know what happens in the warehouse decide about workers' future. "You can't even get near them," he concluded. And yet Castel San Giovanni and Seattle are connected by a dense web of linkages that allow the circulation of things, people, data, and, of course, money. Because of its peripheral position, the warehouse needs an immense amount of infrastructure. Not only a hard infrastructure made of asphalt, concrete, and fiber, but also a "soft" infrastructure made of code and data. Taken together, these hard and soft infrastructures serve as what Ned Rossiter calls "logistical media":

technologies that coordinate and control the global movements logistics is based upon.[22] They function as a logistics operating system that allows commodities to be moved efficiently. Bezos himself has said that the two cannot be disentangled. In his words, "fulfillment by Amazon is a set of web services API that turns [a] fulfillment center into a gigantic and sophisticated computer peripheral," connected in turn to a broader logistics system.[23]

This large operating system is relentlessly expanding. Beyond Seattle, the company has major headquarters and campuses in other cities, including Luxembourg in Europe and Hyderabad in India; offices and data centers in three continents; and a massive network of thousands of warehouses in Europe, North America, Asia, with planned expansions into South America. The FCs which serve as this operating system's main peripherals are typically huge suburban warehouses staffed by workforces that vary between roughly 1,000 and 5,000 employees, depending on the size of the FC, the degree of robotization, and season. FCs are named after the main international airport in the area. For instance, the name of the Castel San Giovanni warehouse is MXP5 because MXP is the airport code for Malpensa, the main hub for the city of Milan. YYZ1, YYZ2, and so forth are the FCs around Toronto, whose main airport, Pearson, is labeled YYZ. SEA8 is near Seattle, BCN1 near Barcelona, EDI1 near Edinburgh, and so on. Fulfillment centers are further classified along other criteria. "Sortable" FCs store items that can be handled by workers, sometimes with a degree of robotization. "Non-sortable" FCs contain bigger commodities, such as bicycles or washing machines, and need to be equipped with specific robots.

Fulfillment centers must be positioned near the metropolitan areas where consumption is concentrated, but also placed far enough from the city center to facilitate expansive spaces and proximity to the airports and roads that allow commodities to be moved around. These hubs are complemented by thousands of smaller warehouses called sortation centers, receive centers, or delivery stations. The latter are positioned closer to customers—inside urban centers or in small cities where there is no FC. Guided by Amazon's algorithms, these smaller peripherals receive inventory or packages from FCs and sort and deliver them to the final customer. The company doesn't only add new FCs in service of geographic expansion. Each time it adds a new warehouse or deposit, Amazon makes its network more dense and more flexible. MXP5, for instance, is now connected to a network of warehouses, including a robotized sortable FC near Rome (FCO1), a non-sortable FC near Vercelli (MXP3), and dozens of smaller warehouses that cover most major urban centers. Many more are being built.

Warehouses employ the majority of Amazon's global workforce, but the human and robotic engagements that move commodities inside FCs are not the only forms of labor that make possible the company's e-commerce operations. Many tasks are required to enable a customer to receive a commodity through an online order or a request to their Alexa. Thousands of engineers and coders work from Amazon's headquarters in downtown Seattle, and in other urban centers too, including Toronto and Milan. And they work alongside hundreds of employees focused on marketing, sales, management, and administration. Alongside the software code running in a global network of data centers filled with racks of computer processors, their labor underpins the functioning of Amazon's e-commerce websites and warehouse processes. And then there's delivery. In many countries, such as the US, UK, or Canada, this is outsourced to drivers employed through the company's gig economy app, Amazon Flex. Customers provide labor too, albeit unpaid, by allowing the company to use the data it generates by monitoring their behavior, for instance, when Alexa records their conversations or when they review products on Amazon.com or Amazon.it.[24] The complexity of such a division of labor is difficult to map out. Amazon itself relies on algorithms to coordinate this global chain, whose links are as geographically dispersed as they are interconnected by flows of data, money, and commodities.[25]

Labor scholar Ursula Huws has defined this global division of labor "fractured." It involves not only core workers hired directly by a corporation—labor that Huws calls "inside the knot"—but also those brought in through outsourcing, working at a distance. And also many who move between these two categories.[26] While Bezos describes the FCs as peripherals, when it comes to labor they are more accurately recognized as the center of the entire company. When we think of work under digital capitalism, we tend to imagine urban, hyper-connected labor, regardless of whether it is coders in San Francisco, food delivery couriers in Berlin, or social media content moderators in Delhi. Piacenza is hardly on the map. But it is an example of the suburban periphery where a new landscape of work is taking shape. This has been in development for a while now. In the 2000s, I was part of a precarious workers' collective that saw the suburban mall as a new outpost of contemporary capitalism. We thought that the mall's mix of consumerism on steroids for shoppers and full-time precarity for workers marked it as a crucial battleground in capital's offensive against labor, and thus crucial as a political target. Far from being what anthropologist Marc Augé had once called "non-places," we saw the mall, the airport, and the outsourced call center as the key sites of a new modernity.[27] Ground zero in the battle for labor.

As e-commerce becomes the dominant form of consumption, the warehouse supplants the mall. And so the warehouse becomes today's frontline of contemporary capitalism, both ideologically, organizationally, and politically. The current battle for the future of work is increasingly being fought in suburban warehouses. Amazon's warehouses in particular.

THE MYTH OF REDEMPTION

Besides the jobs, trucks and concrete, what Amazon brought to Piacenza and to the dozens of other suburban areas which host its warehouses is a myth: a promise of modernization, economic development, and even individual emancipation that stems from the "disruptive" nature of a company heavily based on the application of new technology to both consumption and work. It is a promise that assumes that the society in question is willing to entrust such ambitions to the gigantic multinational corporations that design, implement, and possess technology.

This myth of digital capitalism is based on a number of elements, including magical origins, heroes, and stories of redemption. Some are by now familiar to everyone: A couple of teenagers tinkering away in a garage can revolutionize or create from scratch an entire industry, generating billions in the process. The garage is an important component of this myth. Here we are not talking about the garages where MXP5 workers park their cars after a ten-hour shift in the warehouse, nor about the garages where Amazon Flex couriers store piles of boxes to be delivered. The innovation garage is the site where individuals unbounded by old habits and funded by venture capital turn simple ideas into marketable digital commodities. Nowhere does this myth

run deeper than in California: William Hewlett and David Pack-
ard's Palo Alto backyard shack is listed on the US National Register of
Historic Places as "the birthplace of Silicon Valley," while the garage of
Steve Jobs' parents' house (where he and Steve Wozniak built the first
batch of Apple computers) has been recently designated as a "historical site" by the city of Los Altos. These garages have even been turned
into informal museums and receive thousands of visitors a year, some
even arriving with organized tour buses. For Californian historian
Mario Biagioli, the garage has become an important rhetorical device
in contemporary discourses, helping mythify the origins of contemporary innovation. Masculine innovation in particular, since the garage
is a strictly male space.[28] Bezos himself started Amazon in a garage,
albeit not in California—or so Amazon's origin myth goes: in 1994
he left his lucrative but dull Wall Street hedge fund job and wrote a
business plan while driving cross-country from New York to Seattle,
where he used his and his family's money to start the company.

The myth of the redemption and success of the hero entrepreneur trickles down to the warehouse, insofar as Amazon presents
work to its employees through the frame of emancipation. The idea
of redemption through work is nothing new. On the contrary, it is a
damnation common to modern society. In the early 1960s, militant
sociologist Romano Alquati pointed out that the culture of mid-20th-
century Italian factories included the construction of a "myth" or
"cult" of emancipation. In this instance, it was directed at the masses of
migrant workers who, following World War II, moved from the rural
south to the north of the country to find manufacturing work with
the flagship companies of the Italian postwar economic boom, such as
FIAT or Olivetti. Redemption from the backwardness of rural life was
ensured not only by steady paychecks and the prospect of a pension
at the end of the line, but also by participation in technologically
advanced production processes—the assembly line of industrial capitalism. Amazon simply repeats and updates such promises. In Italy, for
example, Amazon positions itself as an employee-focused company
that brings stable employment back to a precarized labor market—a
boon to a labor market hit by financial crises, lackluster growth, and
lack of opportunities for retraining and upskilling. So Amazon continues a historical trajectory of Italian capitalism, but imports onto

the local context novel characteristics borrowed from the American digital corporation model.

Indeed, digital capitalism updates industrial capitalism's promise of economic and social emancipation with some novel elements of its own. Rather than simply swapping out the assembly line with the robot or the algorithm, the culture of digital capitalism mixes libertarian ideology with entrepreneurial elements. At the core of this myth lies a form of individualism. The combination of new information technologies with free-market dynamics enables emancipatory potential for the entrepreneur.[29] Furthermore, digital capitalist companies state that they exist to change the world, to make people happy, to create value for everyone and not just for investors—technological optimism at its apex.[30] After all, how could you deliver a bad outcome when your first principle is *don't be evil*, as Google's old slogan famously put it.

Amazon extends this old myth to all its workers. Indeed, in corporate documents, the company goes so far as to state that everyone is an "owner" at Amazon. While this is quite literal in the case of engineers and executives who receive shares of the company, it can only be understood at the level of mythology for warehouse workers. A figurative or spiritual commitment to the company's destiny. Managerial techniques used in the warehouse contribute to building this myth, as associates are asked to have fun at work and help Amazon *make history*, as one of its corporate slogans goes. The myth brings with it the idea that there is no alternative to digital capitalism. Only co-option, or failure for those who can't keep up or won't adapt or submit.

Myths are not just old stories or false beliefs. They are ideas that help us make sense of the world. The myth of digital capitalism itself is not simply fictitious, but instead has very concrete effects. For Big Tech corporations, this myth projects a positive contribution to the world, helping to attract workers and investment, and boost corporate value on financial markets. But it has other concrete effects as well. In different areas of the world, and in different communities, the myth of redemption stemming from participation in high-tech production has impacted economies and cultures. Feminist media studies scholar Lisa Nakamura recounted how, in the 1970s, electronics manufacturers operating on Navajo land in New Mexico justified the employment of indigenous women. Labor in microchip production was presented as empowering for the crafty and docile Navajo women—assumptions derived from racist stereotyping.[31] Italy is completely different from the Navajo Nation, and yet the idea that an imported version of American digital capitalism can be a force for collective modernization and individual emancipation is alive and well there too. Belief in this myth is evidenced in many different and even contrasting ways. Some bring resources, like the $1.5 billion state-owned venture capital fund launched in 2020 by the Italian government to support start-up companies in the hope they will foster economic growth. Others sell resources off, like when mayors of small towns with high unemployment compete to attract the next Amazon FC, offering the company both farmland newly opened up for development and a local workforce ready to staff the warehouse. Over the years, the mayors of Castel San Giovanni have described the presence of MXP5 as a force of "development" and a source of "pride" for the town. This is not unique to Italy. American mayors are routinely quoted praising the arrival of a new Amazon facility as a "wonderful" or "monumental" thing for their town.[32]

Amazon's corporate slogans also hedge up its myth. Central is the valorization of disruption—the idea of a hero entrepreneur defeating the gods of the past. Some of the slogans (the so-called Leadership Principles) are repeated time and again and painted everywhere in the warehouse. While Aboutamazon.com, the company's corporate website, describes them as "more than inspirational wall hangings," that is exactly what they sound like. *Customer obsession* is perhaps the most famous one, a slogan that captures the strategic goal of focusing on customers' needs: the rest (profits, power) will follow. It also signals that workers are by design an afterthought. Other slogans are even more predictable, like *Leaders are right a lot* or *Think big*.

Amazon's myth trickles down to fulfillment centers like MXP5 in many ways. Amazon routinely conducts marketing operations aimed at finding new workers, not new customers. Billboards sporting smiling warehouse workers, recruitment events, and glowing articles commissioned by staffing agencies in the local newspaper are common sights in Piacenza, as in the areas surrounding other FCs. Social media multiplies the message. Amazon encourages employees to join its army of "ambassadors"—workers who plaster social media with positive stories about their job or videos in which they happily dance inside the warehouse. Like the FC's walls, all these practices are soaked with the

Leadership Principles: at a recruitment event near Toronto, slogans, such as *Fulfilling the customer promise,* were projected as part of a slideshow filled with smiling arrow logos, accompanying a presentation of more mundane details like job descriptions or benefits. "Every Amazonian who wants to be a leader," we were told, should focus on "customer obsession" and "never settle," and let's not forget that Amazonians "are right a lot." The event wrapped up with free pizza.

AMAZON'S RESERVE ARMY

Top-down myths like those advanced by Amazon are not always accepted uncritically. Myths are malleable and plastic and can be challenged. The reality on the ground in Piacenza and elsewhere has led many to question Amazon's promises of emancipation and modernization. Take, for example, Southern California's Inland Empire. Today, Amazon employs about 20,000 workers in the region, and while unemployment has dropped since Amazon's arrival, the share of people living in poverty has increased. In the US, journalists and scholars alike have reported how many Amazon workers rely on food stamps to make ends meet, and after the opening of a new fulfillment center, household incomes in the surrounding area tend to drop. In 2018, an Economic Policy Institute report titled *Unfulfilled promises* showed that most Amazon FCs create jobs in warehousing but do not lead to an overall growth in local private-sector employment, as many other jobs are lost.[33] The warehouse tends to monopolize employment: "It's either Amazon or nothing" is a common sentiment expressed by FC workers globally. Even *The Economist,* based on data from the US, claimed that Amazon warehouses do not boost wages for workers in warehousing. Amazon, the magazine suggested, can pay less than its competitors because it employs young and inexperienced workers with minimal qualifications.[34]

Back in Castel San Giovanni, MXP5's influence is evident. It has become by far the most important employer in town, if not in the entire province. The warehouse's doors are open to anyone who is looking for a job and is in possession of now-common skills: Marx would have called it the "reserve army" of capital. In his time, industries needed workers who could show up on time, follow the schedule, and respect the employer's property—it took mass enrollment in schools to train generations of workers to follow the rhythms of the factory rather than the cycles of farming. The basic skills required of today's workers have changed, becoming much more sophisticated. But modern education obliges. Huws describes the reserve army of the digital economy as "a plentiful supply of computer-literate workers who can be taken on when needed and dropped when they are no longer required."[35] Local high schools in Piacenza offer certificates for logistics technicians, but those higher skills are not what the mass of associates need.

When I went through the selection process to become a seasonal associate through the local staffing agency office in Castel San Giovanni, I was asked to complete simple tests with about ten other applicants. Tasks involved the recognition of colors and shapes, and decisions about whether to share information with peers—there were fewer copies of the instructions than people in that room. I also went through a quick interview with two Adecco staff, but in other countries the process is even simpler. In Toronto, for instance, we were told there were no one-on-one interviews for fulfillment positions. All that was needed was the completion of an online personality test (many FC job ads mention a "positive attitude towards work" as an asset), a high school diploma, and, of course, the ability to lift heavy weights and stand or walk for 10–12 hours a day.

In the aftermath of the 2008 financial crisis, it was not difficult for Amazon to find a plentiful supply of people to staff its newly opened warehouse, MXP1—the larger and more technologically advanced MXP5 would later be built on the other side of the highway. The crisis had generated a big local mass of unemployed people willing to take up a minimum-wage job. It is then that some of my friends started working there, people in their 30s or 40s who had lost their jobs during the crisis and certainly had a positive attitude about working in a multinational Big Tech corporation offering full-time contracts.

With the crisis also came relaxed labor laws, ready for Amazon to exploit. In fact, not all workers come to Amazon on equal terms: perhaps the biggest distinction is between the seasonal, temporary and the full-time workers. To fulfill the needs of a market that functions according to year-round cycles, Amazon adopts a dual employment model across its global network of FCs: it has a core group of workers employed directly by the company, and a flexible workforce provided by staffing agencies, who can be hired en masse in the Fall and let go in January after the main seasonal peak of work has passed. The position of a full-time associate can be appealing, as it comes with the potential for desirable benefits. In Italy, for example, labor rights won in the 1960s ensure that workers hired directly by Amazon access a national contract that provides minimum wage and steady paychecks, contributions to the national pension fund, six weeks' vacation per year, and a bonus worth one month's salary in December. On top of that, they cannot be fired without cause. In other countries, like the United States, some of these benefits are lacking: there is little protection against job loss, and forget the extra salary in December. But Amazon offers benefits such as health insurance, two days of paid time off, and a $15 per hour baseline salary.

On the contrary, the position of temporary workers is precarious to the max, as they are hired to cope with peaks around Prime Day or Christmas, when orders balloon. These workers have little to no job security. At MXP5, the provision of temp workers is outsourced to major multinational agencies such as Adecco and Manpower. In some countries, public institutions are involved too, as in the case of Germany's *Arbeitsamt* (Job Centers).[36] The contracts of these workers may last as little as weeks, and even the hours of work are uncertain. In Italy, their precarity is the result of political choices, such as the introduction of "MOG" contracts (*monte ore garantito*) that provide a baseline of guaranteed hours of work, say ten per week, for a minimum of one month. Workers can be asked to put in more time but only as decided by the company and with a 24-hour notice. Adecco itself markets this contract as providing "cost reduction with the ability to use labor power only when needed" and "great elasticity"—for firms, of course.[37]

Workers, regardless of their employment status, are also able to take on substantial amounts of overtime, especially during seasonal peaks, when MXP5 runs at top speed. Piacenza is still a rural province, meaning seasonal workers have more historically loaded up hours in this way by working night shifts at one of the many local canned tomato factories, processing the product we still call "red gold." The produce needs to be processed 24/7 in July and August to keep up with ripening, making seasonal work plentiful. Today, logistics has eclipsed food processing when it comes to the number of people it employs, but it serves a similar role. Many of the young seasonal associates I met while researching this book expressed appreciation for the money that comes with the countless overtime hours they put in at MXP5 around Prime Day or Christmas—and may still can tomato in the summer.

In both industries, the composition of the workforce has been shifting dramatically since the early 2010s. When Amazon first arrived in Piacenza, the wide majority of workers were people like me—white Italians born and bred in the region. But today's warehouse is staffed by a highly diverse population. Amazon has all but exhausted the local population drawn from both the infinite series of small towns that punctuate Piacenza and the nearby provinces of Pavia and Lodi. As I was told by Peppino, who lived in one such small town, "everyone from around here has worked at Amazon or in one of the other warehouses. Everyone." And yet fulfillment requires bodies, a continuous flux of fresh bodies. During production peaks, the company cannot rely on the local workforce to sustain shifts that can require up to 3,000 workers, about twice as many as the full-time associates who work at the warehouse year-round. Every year, hundreds of thousands of temp workers are hired across the world. To catch up with its need for flexibility, the company has enlisted more workers in its reserve army by looking beyond Piacenza and incorporating migrant labor. During peak seasonal periods, unbranded "Amazon buses" run by temp agencies drive dozens of precarious workers from neighboring cities like Alessandria or Parma, and from suburban working-class neighborhoods in Milan (one hour away from MXP5) to work peak shifts.

In their early days, industrial capitalism and manufacturing relied upon labor from masses of assembly line workers. They both tended to machinery and engaged in manual labor, taking up processes that could not be mechanized. In the US, factories in the industrialized East coast and Midwest replenished this workforce from the continuous influx of migrants from Eastern and Southern Europe, and the arrival of freed Blacks from the South. In Italy, factories in the industrialized north benefited from internal migrations, such as those of peasants from Southern regions in the 1950s and 1960s.

Today's workforce composition is different, but some dynamics overlap with the old days of industrial capitalism. For instance, Amazon's fast expansion means it needs to onboard masses of new workers every year. Its employees thus tend to be young: data from the US Census Bureau suggest that nearly half of its American warehouse employees are under 35. These new younger workers tend to be members of racialized minorities. Data from the company shows that as of late 2020, Black and Latinx workers were overrepresented in Amazon's workforce, accounting respectively for 26% and 22% of the total. At about one-third of the total, white workers are underrepresented in the entire workforce, but the picture flips when it comes to management, where Amazon reports they occupy 56% of the positions. Power is also more commonly in the hands of men, who represent over 70% of Amazon's managers globally.[38] These numbers are consistent with the unequal distribution of power and money in contemporary racial and patriarchal capitalism, of which Amazon is a bastion.

The racial nature of Amazon labor is visible in Piacenza too. You only need to drive by the local Adecco chapter to see the young workers of color who wait on the sidewalk for their turn to apply for a job at MXP5. Many are migrating for the second or third time, as they relocate from other parts of the country to Piacenza for Amazon. More than once, friends from Southern or Central Italy who knew I was working on this book asked me for help finding housing for new MXP5 workers who were moving to the area. For instance, someone from Apulia in Southern Italy texted me to ask for help finding a home "for my friend from Senegal who is about to be hired by Amazon in Castel San Giovanni [...] He couldn't find any new job in Bari [...] so he spent a couple of months in Veneto as a farm worker and now got a job at Amazon!" Amazon FCs are such powerful attractors that it has become common to hear stories about workers living in their RVs in the warehouse's parking lot during seasonal peaks—and being let go in January once the peak is over.[39]

This demographic push is not without consequences for local politics. Migrant workers from the Maghreb and young women form the backbone of SI Cobas, the militant union that organizes most warehouses in the local logistics industry, with the notable exception of MXP5. But while it can prove fruitful for some unions because it brings in new members ready to take up the fight, this demographic composition does not necessarily bode well with the white Italian inhabitants. Castel San Giovanni has one the highest migrant populations in the region, and also one of the most right-wing electorates. Here, in the 2019 European elections, the xenophobic far-right party Northern League, whose racist anti-immigrant rhetoric has a strong grip on the precarized white working and middle class, received over 50% of the votes. The city used to be significantly more progressive, back when the Communist Party still ruled the Emilia-Romagna region where Piacenza is located. Before the Berlin Wall came down, a song from legendary 1980s punk/new wave band CCCP called Emilia "province of two empires," its lifestyle influenced by America and its politics and economy connected to the Soviet Union. Now Amazon links the local economy to the US, and is threatening to deal the last blow to the remnants of the past era, such as the Ipercoop mega-malls federated with a different league, the League of Cooperatives.[40]

After all, many things have changed since the fall of the Wall. The hyper-precarious MOG contract used by MXP5's staffing agencies was introduced in tandem with a package of reforms that made it easier to fire even full-time employees. Known as the *Jobs Act*, the reform was sponsored by the fiercely anti-labor government of Prime Minister Matteo Renzi, then secretary of the center-left Democratic Party, and Labor Minister Giuliano Poletti, the former President of the League of Cooperatives. It could have been worse: in Spain and the United Kingdom, workers have reported being hired through a zero-hour contract that asked them to be available without any guarantee that they would even have the opportunity to work and earn a wage. Amazon exploits and pushes the boundaries of local labor laws to fuel its global system of precarious work.

THE WAREHOUSE IS THE NEW FACTORY

Most of the workers who enter the gates of the gigantic fulfillment center every day have never worked in a factory. Nevertheless, many compare MXP5 to a sweatshop and describe the work as assembly line labor. Factory comparisons are what first came to mind for me as well, the first time I ventured to MXP5's parking lot. The flow of dozens of young workers in and out the gates of the warehouse between shifts instantly reminded one of the masses of workers walking into a factory. This is simply a personal impression, albeit shared by many. But it also resonates with a broader reality: the Amazon warehouse does indeed incorporate and renew some of the dynamics of the industrial capitalism of yore. In this sense, Amazon continues and extends a process that began with the industrial revolution. But insofar as the warehouse can be likened to a factory, it is a digital one, a fruit produced from the grafting of contemporary logics onto the trunk of industrial capitalism. In this tension between the old and the new, Amazon adds futuristic technology to its arsenal of tools for organizing labor, and at the same time reproduces old-fashioned ways of controlling the workforce. In a sense, it is the digital version of the tumultuous heyday of early industrial capitalism.

For German sociologist Moritz Altenried, we are witnessing the emergence of a digital factory where new forms of automation are responsible for inserting human labor in production processes based on machines, and of course for extracting value from it.[41] At Amazon, the assembly line has been replaced by the algorithmic organization of the labor process, but both are used to standardize tasks, optimize processes, and reduce the time needed to train new workers. Robots and software systems intensify labor and make it more dangerous rather than facilitating it. Like in a factory, workers must be convinced and motivated, and to this end Amazon relies on a despotic and paternalistic workplace environment. But management is augmented by the use of digital surveillance to monitor labor and control workers' performance, and by organizational techniques built upon the myth of progress offered by the high-tech corporation. The clockwork has been replaced by the algorithm, but workers still have to synchronize with the rhythms of work dictated by machinery. Like early industrial capitalism, Amazon relies upon a highly precarious workforce that can be onboarded and discarded at will, and at times must be even bused in. But Amazon plans their obsolescence more carefully, as it encourages (or forces) workers to quit the warehouse in ever-faster cycles. In its projects for the warehouse of the future, Amazon imagines and desires a workplace where these trends are expanded and where human labor is even more subordinate to machines.

In a very different Italy from the one Amazon operates in today—the economically booming Italy of the early 1960s—Italian theorists of *operaismo* (workerism) such as Romano Alquati, Mario Tronti, and Raniero Panzieri set out to understand the transformations of labor and the evolving relations between workers, capital, and technology.[42] My own work is indebted to that story. When I started researching labor in the logistics district around Piacenza I noticed many similarities to what this small group of intellectuals had identified in factories.

For instance, they thought that the industrial working class' central role in the evolution of capitalism was being overlooked, and thus its revolutionary potential was being overlooked as well. At the core of their analysis was capital's struggle for control. Capital, they thought, must control workers' natural unrest and mitigate their refusal to cooperate with corporate goals. For this reason, capital had a "plan," and technology was a key component, although the workerists saw labor, not capital, as the real engine of change. It's an old tension resurfacing in novel ways. Warehouse workers demand political changes: a reduction of flexibility and work rhythms; a just, healthy, and safe workplace; and a redistribution of the immense profits accumulated by Amazon; among other things. The corporation offers technological fixes, from wellness apps to AI-powered social distancing cameras, and union-busting techniques to prevent worker organizing.

Rhetoric about technology's potential to enact radical change, or the myth of digital capitalism's disruptive potential, fails to acknowledge that change in the workplace is a political process. None of this could be possible without the decades of precarization, diminishment of workers' power, expansion of globalization processes, and rise of the financial market that have made digital capitalism possible in the first place. Theorist Ruha Benjamin says that technology is but one of the factors that contribute to capital's ability to "innovate inequity."[43] The development and application of new technology is part of a broader project, at least under capitalist relations of production in which it is designed and used by capital itself. Amazon's technological power would be nothing without its economic and political power—and without its relentless drive to accumulate capital. It is also on a collision course with workers and their communities: how long will it be until Amazon's empire collapses?

MXP5 may be just a small cog in the company's transnational money-making machine, and yet like other fulfillment centers it can be used as a lens to understand the company and its role in the evolution of contemporary capitalism. Such knowledge can only be built in concert with workers, who alone experience the organization of labor under Amazon and imagine forms of resistance to it. They work in a single warehouse but are connected to the rest of the Amazon workforce through national and international worker-led organizations. They experience standardized processes and managerial techniques imported from the United States, as Amazon replicates them everywhere in its global network of warehouses. They encounter other associates from around the world in online spaces where they can find each other and overcome the distance created by the transnational nature of the company and by managerial control inside each individual warehouse. Like the factory, the warehouse is not isolated from society. Quite the opposite: its logics are expanding to other spheres of our lives, other jobs and other industries. But to understand Amazon's impact on work and imagine an alternative to its plan, we must first descend into it, this new factory of digital capitalism, cross its gates, and meet those who work in the warehouse.

2

Work hard

Work hard. Have fun. Make history. This slogan adorns the interiors of each and every Amazon fulfillment center. It is the first thing you encounter as you walk past the main door into MXP5, painted in the entryway, right before the security gates and the body scanners all workers have to pass through to enter or leave the warehouse. And it appears on many other walls inside too. The first part of the slogan, *work hard*, is certainly something most workers are prepared for. It is no surprise to anyone that Amazon work is difficult, fast, and demanding. Everything inside the warehouse is in service of the speed and efficiency Amazon promises to consumers, from the algorithms used to record the position of inventory to the night shifts and overtime hours required of workers. Indeed, it takes a lot of hard work to store, retrieve, pack, and ship the hundreds of thousands of items that enter and leave the warehouse every day. For first-time visitors, the size of the facilities themselves is astonishing. MXP5 associates sometimes refer to their workplace as "the spaceship" because of its resemblance to the engine room of a science fiction ship: a gigantic, windowless space, illuminated by neon lights and criss-crossed by miles of conveyor belts moving items and boxes from one area to another. The clean, sanitized environment is demarcated by yellow and blue lines painted on the floor, guiding people's movements. Yellow metal staircases lead up to the heart of the warehouse: a central multi-floor area that Amazon calls the "pick tower."

As far as you can see, the massive floors stacked on top of one another of the pick tower are lined with thousands of shelves, each divided into colorful, gridded compartments, packed with the goods Amazon sells through its websites. That is where I first encountered the immense accumulation of commodities stored in the warehouse: books, cat litter, toys, office supplies—and really every sort of item you might expect to find somewhere in a large mall—are crammed into these cells. Just as the shipping container is the standard object used in global logistics chains to move commodities across the world, a bright yellow bin is the standard object Amazon uses to move commodities in and out of the pick tower. Hundreds move in all directions at any given time, on carts pulled by workers and on the automated lines carrying things to the packing and shipping departments. The white noise produced by the conveyor belts serve as a soundtrack for this scene, layered under the music blasted by huge speakers in some areas of the warehouse. But in the pick tower, as in a library, the dominant sound is the silence of the dark aisles punctuated by the quick steps of workers walking to reach whichever shelf contains the product they have been assigned to retrieve.

As in the rest of the warehouse, the work done inside the pick tower is dictated by technology. Obviously, the mere existence of e-commerce is premised on digital technology: the websites, computers, and phones without which no order could be placed online. Technology is also used to prop up the efficiency of the company's operations: the chain of events that brings a package to your doorstep the day after placing an order is made possible by a complex system of algorithms that know where a commodity is stored and can assign a worker to retrieve it, all while directing others to pack, ship, and deliver it. Amazon refers to this system as Mechanical Sensei. There are more technologically advanced workplaces out there. Actually, MXP5 is not even the most technologically advanced workplace in Piacenza: alongside the Zalando, IKEA, and other Amazon-type warehouses that compose its logistics hub, the province hosts a small but sophisticated mechatronics industry. Yet, at Amazon technology occupies a special role. And how could it not: Amazon presents itself as a tech company. Its technology is the subject of endless company boasts, bevies of newspaper articles, and rapt consumer engagement. And it is often imagined as a workplace that will soon be fully automated by robots.

This fixation on technology often serves to push workers into the background. But for their part, workers know very well that the warehouse needs their living labor: they are the real engine behind Amazon. As one manager told me, "technology codifies, understands, and manages. But the real machine is the human: everything is done manually." The two forces cannot be unyoked. To maintain its promises of increasingly fast delivery, Amazon must use technology to boost workers' productivity (that is to say: speed up their labor); turn commodities into information so that they can be managed by software systems; standardize tasks so that any employee can perform them; facilitate worker turnover in case there is a shortage of labor; and strictly control workers by reducing their agency and giving more power to management. Technology, in sum, makes fulfillment and delivery possible—not only by enabling logistics, but also by enabling Amazon's control over its workers.

By using information to optimize labor, Amazon is continuing a long-standing trend of modern capitalism. In the 1880s, American engineer Frederick Winslow Taylor applied a scientific approach to the management of workers and machinery on the shop floor. His "scientific management" was based on the idea that one could carefully analyze the labor process to find ways to shave time off tasks. In practice, this meant supervisors roaming the factory, armed with stopwatches and notebooks, recording workers' activities to inform new protocols. Workers were naturally lazy, Taylor thought: an idea shared by Jeff Bezos himself.[1] Thus management could use his "time–motion analysis" to figure out things like the most efficient movement to perform a particular task, for instance, tighten a bolt, and thus speed up their labor. Taylorism, as it became known, soon grew into a common management tool, influencing factories the world over. A great way to make people work faster, harder.

And the model is still with us, as many corporations now use digital technology to renew Taylor's techniques. In his research on MXP5, sociologist Francesco Massimo talks about the "specter of Taylorism" hovering over the warehouse.[2] In Amazon's digital factory, the supervisors' stopwatches and notebooks are replaced by the digital analysis of data generated from human labor. This analysis is then applied downstream to optimize and control that labor. The main instrument of both analysis and control is the barcode scanner, which workers commonly refer to as a "gun." Most are handheld wireless scanners— the same technology used by supermarket cashiers to scan prices—but these devices can also be found mounted onto wristbands strapped to the worker's hand or attached to workstation computers. The workers who staff the warehouse pick a scanner up at the beginning of their shift from huge charging stations: entire walls filled with dozens of scanners set to recharge. Their first task upon beginning a shift is to scan (or "shoot") the barcode on their badge, thus logging into the system. Through the barcode, workers are rendered into information, too—just like the objects they work with. This fact is not lost on them. A veteran of MXP5 (fulfillment centers are also coded) named Maria emphasized this point to me over coffee in Castel San Giovanni: "We are a number; what matters is our badge, our code," she told me. "You saw it: everything at Amazon is a barcode, and that's what we are too, and it's sad." Indeed, I did see it. Amazon organizes guided tours of its fulfillment centers, and every time I walked into one, be it MXP5 in Piacenza or others I visited elsewhere in Italy and Canada, the employees tasked with showing the small crowd of visitors around described barcode scanners as the main instrument through which Amazon connects customers, commodities, and workers.

Upon picking up their barcode scanner or logging into their workstation, FC associates are caught up in a form of work made possible by Amazon's technological infrastructure. The worker's scanner begins mediating between them and management: the software that runs it breaks down complex processes into individual tasks that can be assigned to any worker in the pick tower, communicates orders, and monitors and optimizes workers' activities, organizing their labor. Most decisions are made by the software systems which crunch data about inventory and workers, rather than by human managers on the warehouse floor. Automation is part of the picture in many other ways, as a lot of robotic work sustains Amazon's operation. Just think of the virtual assistant Alexa and how engineers built a precise kind of womanhood into her, so that she can better serve its owners: subservient and always available to turn their orders into commodities delivered from the warehouse to their homes.[3] But as important as the technology may be, the physical and repetitive hard labor of the warehouse relies on masses of workers. Without them, Amazon would quickly grind to a stop: as in industrial capitalism, the technology used in digital capitalism means nothing without the mobilization of human labor at a massive scale. In many ways "automation is very limited," as Maria put it. "The real automation, the real plus, are the algorithms that organize the customer's order, group items based on what and when they have ordered, whether it is available or must come from another FC, that's the system's intelligence." But, she pointed out, that sort of automation relies on human workers. "Do you know what's the main resource inside the FC? Us, labor, our arms."

FOLLOW THE COMMODITY

Passing through the fulfillment center, each and every commodity sold by Amazon goes through four core processes: receive, stow, pick, and pack. All involve both machinery and human labor. Let's consider the trajectory of a pink coffee mug. The first two processes it will encounter in the warehouse are part of "inbound" work. At the "receive" work-stations, laborers pull boxes of commodities off of incoming pallets. Opening these boxes, they find our mug alongside many others like it, each identified by a unique barcode. By scanning this code, workers record the arrival of the mug in the warehouse. Next, the products are put onto a conveyor belt, where they travel to the "stow" area. Workers there group them into yellow bins, which are loaded onto carts, ready for the pick tower. At this point, workers called "stowers" grab a cart, scan its bins to record that they are in charge of storing its content on the shelves, and walk to their assigned area of the pick tower. As they place the mugs on the shelves, they use their barcode gun to scan both the code of the mug and that of the cell where they have stored it. The inventory system now knows the position of each copy of the mug in the pick tower.

From here, other workers engage in two types of "outbound" work: "picking" and "packing." Let's say you are ordering the pink coffee mug to Bologna from Amazon.it. The software system searches the inventory to figure out which fulfillment center serving the city has a copy of the mug. In all likelihood, this will be MXP5. The system then assigns the task of retrieving it to workers called "pickers," causing their device to give them the item's location. They then walk into the storage areas of the warehouse, retrieve the items assigned to them by the inventory system, and carry them to sorting workstations. A picker may be assigned different commodities that are to be shipped to different clients, and if an order contains multiple items (perhaps with the mug you also ordered a USB drive and some eyeliner) those are retrieved by different pickers. Once the items are dropped off at the sorting station, other workers sort the items by scanning them and placing them into bins that correspond with individual orders. These bins are then loaded onto a conveyor belt and moved to the "packing" stations. There, the system tells a worker called a "packer" the appropriate size of the cardboard box or envelope they must use to wrap up the order. In seconds, the packer transfers items from the bin to the prepared package, which is sealed and loaded onto the conveyor belt, where it will be scanned by a machine that automatically generates, prints out, and pastes onto the box an adhesive tag containing the customer's address and yet another barcode. All this information is machine-readable only—no worker can see who has ordered what. Moving further along the conveyor, the package passes under another scanner that reads its order barcode and sends it automatically to a designated bin, where it will be picked up by a particular carrier, as determined by geographical delivery area or shipping priority. Workers load orders onto pallets and then onto trucks waiting at docking stations, which will bring the packages to sortation centers: smaller warehouses near or inside the destination city or neighborhood. From there, your package will be assigned to a driver and finally delivered to your home.

Other things happen in the warehouses too, beyond these core processes. For instance, "reverse logistics" workers deal with orders returned by customers (you decided that you don't need a pink mug after all). These workers examine and send returned goods back to the pick tower or redirect them to other fulfillment centers. In "quality control," workers check processes performed by others, ensuring items are positioned correctly on the shelves, for instance. Workers at docking stations load or unload the hundreds of trucks that every day bring items to the warehouse or collect orders.

Common to all these processes is the use of automation in concert with human labor. Amazon's algorithms make the decisions that activate workers, moving the mug through its trajectory from a truck to the pick tower and finally to your home. Which FC should the mug be retrieved from? Algorithms make the call. Which worker is to be tasked with picking, rebinning, or packing it? Again, the software decides. This automation of functions traditionally assigned to human managers is certainly not unique to Amazon. Industries ranging from social media to ride-hailing are based on forms of algorithmic decision-making. Whether it's a newsroom's social media manager working to increase engagement with an article posted on the newspaper's website, or an Uber driver directed by the company's app to an optimally positioned rider, data-driven algorithmic processes control and shape worker activities.[4] The difference is that in the warehouse it is the scanner's screen that mediates between these automated decisions and the worker, rather than a phone app or a browser window.

The use of software systems to organize labor is not a neutral endeavor. Technology in the workplace reflects and maintains the power relations that underpin the employment relationship. This is to say that capital designs and applies technology with its own goals in mind: in Amazon's case, a fast and seamless order fulfillment, and the control of workers. Sociologist Aneesh Aneesh coined the term "algocracy" to emphasize these asymmetrical organizational forms that algorithms allow management to build, and control, to its benefits.[5] For instance, capital has the power to monopolize knowledge about these technical systems. It is not by chance that the inner workings of the corporate algorithms that sustain warehouse processes are opaque and thus difficult to understand. Industrial secrecy and nondisclosure agreements make it impossible to access the code of Amazon's software. But these technologies do not entirely resist interpretation. Automation is a sociotechnical system that affects and is affected by specific people and processes, and thus can be understood in the context of its application in the workplace. Workers themselves also have limited access to these software systems. But because they experience first-hand the effects of algorithms in fulfillment processes, their knowledge can be precious.[6] Not only can they offer their sustained observations, but they can also probe and test on the ground the procedures of data extraction and analytics that organize their labor, for example, trying to figure out the way in which orders are assigned to a certain picker and how that affects one's productivity.

So in order to understand the use of digital technology at Amazon, we need to turn to warehouse workers. They experience something that is inherent in all capitalist relations of production: the tendency to incorporate their labor into machinery. In an industrial era dominated by the steam engine, workers provided physical input, for example, feeding raw materials to machinery. The same basic dynamic is true today. But it's not all about their muscles. Obviously, workers contribute physical labor at Amazon, for instance, by moving commodities around the warehouse, taping boxes of orders, or unloading trucks. They also generate information that software systems capture through the scanner and use to manage the inventory in the pick tower. Software systems turn this human activity into data, feeding it back into complex machinic systems dominated by algorithmic and robotic technology. This is not entirely new. Already back in the early 1960s, workerist theorist Romano Alquati said that workers are producers not only of commodities, but also of information—and thus value—which capital must expropriate and control.[7] But the tight feedback loop that automated, algorithmic systems enable is unprecedented. The novelty, in a digital-intensive warehouse like Amazon's, is that the information generated by workers is algorithmically crunched to make possible and improve the machinic processes that underpin fulfillment—and control workers' own activities.[8] Stowing and picking are the two processes that best exemplify this dynamic. As workers store commodities on the shelves, the result of a complex activity that relies on their individual choices and dexterity is turned into digital data and incorporated into machinery. Once digitized and captured by its software systems, Amazon uses such information to strictly direct the labor of pickers, ensuring that they work in an efficient and controllable manner in fulfillment processes while guided by automated software systems.

FROM CHAOS TO ORDER

I met Mark on a sunny spring day, at a bar patio where many young seasonal associates from Piacenza hang out. An MXP5 temp worker, Mark had been working as a stower for a few months, experiencing first-hand the organization of inventory in the warehouse. His days were spent walking the pick tower, quickly emptying cart after cart by placing commodities on the shelves. It was a monotonous activity, he told me, but he did not mind it too much. He did not have to focus to perform it, but also did not retain much information from it: "As a stower you cannot control and see where you have put a certain object," he said. "You [would] really need to pay attention to remember. There is so much different stuff in the warehouse, and you just have to keep unloading carts." Indeed, while workers stow the stuff, they do not possess an overview of where the stuff is. Such knowledge is only maintained by the software systems that control work in the warehouse. Amazon relies on a computer system able to capture workers' activities and carefully record the location of items into inventory for later directed retrieval. A stower might know where the items they've placed are, for a time. But that knowledge never remains current for long. Inventory in the pick tower is based on so-called "organized disorder" or "chaotic storage" logics. Instead of storing items by category on dedicated shelves—phone covers here, toilet paper rolls there—Amazon stores them in seemingly random locations, and any given cell may contain any combination of different items. As workers place commodities in this chaotic fashion, their labor is immediately captured by their scanner, turned into data, and used to make sense of inventory when something needs to be retrieved. Only software can re-order the chaos. For the worker, it would be impossible to find something in the pick tower without the aid of the scanner and the software that runs it.

In practice, chaotic storage works like this: dozens of workers are assigned bins containing recently arrived items. They are then sent into a specific section of the pick tower, placing items as they go, anywhere they will fit. Crucially, they also record the stowing locations with their scanners. In this way, the stowing of commodities is left at least partially to workers' autonomous decisions. Some types of products, like costly electronics, have designated shelves or areas of the pick tower where they are concentrated. But in most instances, it is random. There is, for example, no designated area a worker needs to walk to in order to stow toys. Instead, these objects are distributed throughout the pick tower. Stowers need only abide by two main principles as they go. First, they are not to stow all copies of a given item in a single area. Instead, they must distribute duplicates throughout the pick tower, in a number of different cells. This practice increases the likelihood that a copy of a given commodity will be close to a picker, thus reducing the time they will spend walking. It is also intended to prevent bottlenecks in the event of an order surge, reducing the likelihood that several pickers simultaneously looking for the same item will converge on the same cell or corridor. Second, they can place items in any cell with enough room to contain them, as long as no commodities of a similar type are already stored in it, and no similar commodity is present in an adjacent cell. Thus, a teddy bear should not be stowed in a cell containing other stuffed animals, but could be placed in one containing a cellphone case, several copies of a textbook, and a T-shirt. This is also aimed at speeding up the future labor of the picker, who will be guided to a location that contains a copy of the teddy bear but no other similar items, thus reducing decision time and also the possibility of costly mistakes.

In a pick tower of several thousand square meters, containing hundreds of shelves holding tens of thousands of cells distributed over a number of different floors, this stowing process generates an inventory that no individual human being could possibly navigate without the aid of Amazon's system algorithm. But human agency remains a crucial component of the system, nonetheless. While stowing is a simple process to comprehend, it relies heavily on workers' dexterity and analytical capacities. The FC manager I spoke with relayed just how complex this seemingly simple process can be: "You receive mixed totes which can contain one copy of an item and one copy of another. In a single tote you can find anything, from a CD to a soccer ball or a book. And then there are smaller and bigger cells." Human analytical abilities are unmatched in their capacity to quickly parse between these items—particularly when the inventory system is wrong. In some cases, the manager told me, humans need to quickly adapt to errors, as when "an item has been cataloged as being smaller than its actual size." Moreover, human creativity is essential to efficient stowing, facilitating the maximization of available storage space in a given location. While Amazon's algorithms can support workers in avoiding wrong choices, human labor alone possesses the flexibility and speed to efficiently store—in a "chaotic" manner—a range of items that are different in shape, weight, volume, color, etc. What workers do not possess, or rather quickly surrender to the machine, is their knowledge of the position of the commodities they have stored. This lack of knowledge is one of the factors that renders them easily replaceable.

Barcode scanners are at the center of chaotic storage. Since commodities and shelves are identified by barcodes, a stower uses a scanner to shoot both the teddy bear and the cell they have stored it in. A green light confirms that the system has recorded the item's position. Once this process is complete, the system algorithm managing the inventory can send a picker to retrieve the teddy bear when an order is placed. This system can be easily subverted, of course. Mark remembered occasions when "I was in a rush so I shot a cell and only then grabbed the item and realized it wouldn't fit in, so I just put it in the cell above." When I asked him if misplacing items could be a form of sabotage, he denied attaching any political meaning to it, but smiled when he added: "I am a sucker for graphic novels and at times I find, say, a cool comic book, pick it up, carry it with me and read a page here and there, and then put it back in the shelves wherever I am at that moment." But Mark would not record the new position where he stored it: hidden in plain sight, the book was now lost forever.

Amazon did not invent chaotic storage. In warehousing, strategies based on this logic date back to the 1970s, when logistics first became computerized.[9] But Amazon most certainly has perfected the technique, which does indeed seem particularly suited to warehouses that store a multitude of different commodities and often fill orders composed of just a few items—or even a single item, as is frequently the case for Amazon. Storing things randomly is efficient. It allows the best use of space on the shelves, as any commodity can be squeezed wherever it fits. It is also useful to meet the tight delivery schedules promised to customers, reducing the time it takes to retrieve a given item—even if it means greater time pressures put on workers expected to pick and ship an incoming customer order the instant it arrives. In fact, chaotic storage increases the probability of always having some items in any given shelving area. In turn, this can reduce the "unproductive walking time" (to use a definition from management literature) spent by pickers who are fulfilling an order.

For information theorist Philip Agre, such forms of "data capture" aim at improving the rational organization of industrial production and services through the tracking of people and objects.[10] For example, a supermarket may track its customers' shopping patterns through loyalty cards and in-store cameras in order to reorganize the way in which it presents products on its shelves. By capturing and controlling the movement of people and things in the warehouse, this Taylorist logic applied to the digital factory model aims at increasing the efficiency of labor.[11] It also has the effect of rendering workers lost in the face of the warehouse's chaotic complexity, and thus completely dependent on inventory software. Even technical literature admits that chaotic storage "makes the orientation of pickers without an information system impossible."[12] This is a key difference from the traditional warehouses where some MXP5 workers were employed before joining Amazon. In the old days, even a carefully organized inventory relied on the memory and familiarity of a stable workforce for top efficiency. By contrast, the environments inhabited by today's warehouse workers are inscrutable to humans and knowable only to the software that orchestrates their work and determines their paths through the shelves.[13] Faced with the impossibility of knowing the inventory in its entirety, humans must outsource all of this understanding to the machine, which in turn entirely determines the organization of the labor process.[14] Workers, in sum, end up depending upon the warehouse and its technological infrastructure, even though their work is essential to create the very conditions of such dependence.

It is worth stressing that the information generated by the datafication and capture of stowers' work is stored and analyzed in computers that are part of Amazon's global system. They may not even be in Italy or in Europe. They are certainly not in the warehouse, nor can MXP5 workers influence or access them.

Thus, chaotic storage and the monopoly it gives Amazon over warehouse inventory information is a form of worker *dispossession by machine*. I use the term dispossession because workers are immediately deprived of a crucial characteristic of past warehouse labor: the requirement for them to "be there" to develop knowledge about the warehouse.[15] A requirement that makes them valuable, even irreplaceable, at times. MXP5 associates who have prior experience in traditional warehouses can sense this difference clearly. At their old jobs they were treated as possessors of valuable knowledge—literally the knowledge of where things were—that was essential to the efficient operation of the warehouse. In this way, they were valued across time. The possession of such a knowledge was power for workers who could rely on it for leverage and job security. Amazon replaces this arrangement with a complex process that involves hundreds of stowers generating a chaotic, algorithmically managed form of inventory that no human being can know in its entirety. So, labor becomes more disposable and easier to squeeze and exploit in subsequent processes like picking.

THE ALGORITHMIC PACE

Once an item has been chaotically stowed, it is ready to be picked, which means that it can be retrieved when a customer orders it. Indeed, every order on Amazon's website triggers a cascade of effects that eventually arrive on the screen of a picker's barcode scanner. Picker is the basic entry-level position for most seasonal associates, but also the job that requires the most employees even when business is slow. It is also the most monotonous and physically demanding. This means it often falls to warehouse rookies—the mostly young, racialized, female workers who represent the bulk of the seasonal workforce. But it is also something many full-timers will do for years on end. These workers are at the receiving end of your order. When you purchase that pink mug, the inventory system uses the scanner to task an associate with retrieving the commodity and sending it to packing and shipping. The task itself is relatively simple: follow the instructions from the barcode gun, walk to the correct aisle in the pick tower, retrieve the commodity, shoot it, and place it in a yellow bin. Repeat. And repeat. And then repeat again for the next eight hours, or maybe ten or more if you're working overtime. The worker is not chosen randomly. Using the information captured during stowing, Amazon's system software knows where each copy of the ordered commodity is located—and thus will aim at choosing the most favorably located picker, in accordance with its calculations. The stowing process has enabled the machine to render commodities into information, and this information is now used to break down the order fulfillment process: no individual worker takes care of an entire order. Like in an assembly line, the process is turned into a series of single standardized tasks. For the pickers, this means: go to aisle X, cell Y, and pick object Z.

Hundreds of pickers walk among the shelves of MXP5's pick tower, pulling a cart that carries the yellow bins they need to fill up with their "batch." A batch might be composed of 50 different objects, each potentially destined for a different order. The objects might include a T-shirt, three books, a sex toy, or a Hello Kitty phone cover. These tasks are communicated through the worker's barcode scanner. It shows the item's name ("red water bottle" or whatever) and image, alongside information regarding the item's position on the shelves and the time the picker has to complete the task—often less than a minute. Zak, a former seasonal picker, described it as a race against time: "the gun shows a countdown for each individual piece, for instance, a minute and a half to go get this item. It's a bar that dwindles. And [the pick tower] is gigantic. You need to walk for 4 minutes to get anywhere." As the worker grabs the item and scans the barcodes on both the item and the shelf location, the system records and approves the process. This is not dissimilar to other forms of algorithmic control that take place outside of the warehouse, where the barcode scanner is replaced by the phone apps commonly used to assign customers to Didi drivers in Beijing or restaurants to Deliveroo couriers in London.[16] In all these examples, some functions typically performed by human managers are outsourced to algorithmic systems: assigning tasks, monitoring workers, scheduling.

At Amazon, this form of algorithmic control is used to speed up work. Pickers are required to keep a fast pace—the so-called "Amazon pace." This is not a run—for safety reasons. Rather, it is the fastest possible walk. Its rhythm is dictated by the scanner, as stressed to me by Zak:

> As you are loading an object onto the cart, the next one appears on the scanner. So as you are loading your cart you start moving, and as you are arriving you already take a look at what you are to pick next, you don't stop, and then you look at the shelf, is it a book or something else? In which area of the shelf is it?

In fact, the algorithmic organization of work in the warehouse is geared toward one main goal: productivity. Work hard, remember? To fulfill its promise of speedy and cheap online consumption—delivery in two days; 24 hours; two hours—Amazon needs to squeeze as much work out of people as possible. In a sense, this is a continuation of another industrial practice: the just-in-time production methods used since the mid-20th century to make manufacturing more flexible and responsive to consumer demand. At Amazon, it is not about producing but about delivering commodities as soon as possible after consumers order them. As the company applies its own version of just-in-time to retail, it accelerates a tendency that has been present in logistics for decades—speeding up and rendering more flexible the transport of commodities across the world, their transit in warehouses, and their delivery to customers.[17]

This may be the main reason why Amazon warehouses have come to epitomize a new form of labor degradation.[18] Many MXP5 associates describe FC work as "a sink or swim job" that is not only physically demanding and repetitive, but also unskilled and alienating. This is true beyond the FC in Piacenza and across Amazon's global network of warehouses. An online review written by an English associate described Amazon work as being "a brain dead activity and tiring at the same time." The reviewer continued: "Imagine walking around for 10 hours and the only thing you do is beep a scanner constantly, pick up random items, and walk along for the next one, rinse/repeat." In the digital factory, supervisors no longer walk around with a stopwatch, measuring (with an eye on reducing) the time it takes a worker to pull a series of levers on the assembly line.[19] The tyranny of the clocks of yore is replaced by today's tyranny of the algorithms that dictate the pace of work. For the picker, this materializes in a continuous stream of tasks, an infinite chase of the next blip on the screen. An ever-faster race against time.

Amazon does not pay its workers based on their productivity: piece work is illegal in most jurisdictions where the company operates. Yet the main generator of difficulty in Amazon work comes from managerial pressure to "make rate" or "meet the targets," which means picking a certain number of items per hour—say 70, or 100. Workers, especially the precarious temp worker whose contract renewal is always hanging by a thread, must just grin and bear it. They have no alternative to the imperative of skipping to the algorithmic beat. At MXP5 and across the hundreds of other Amazon fulfillment centers across the globe, they know all too well what a Texan associate described as the constant "walking, lifting, moving, stepping, twisting, turning your body." Meaning workers are "physically moving non-stop from the start of your shift until your first break" to meet productivity rates. The combination of the rates and the physical nature of the job makes working at Amazon "emotionally and physically draining," "hard on your body," and also "mentally" hard, creating "permanent aches and pains" as well as "anxiety" or "fatigue and depression," to use a mix of expressions from American FC workers. Most pickers identify the characteristic hustle of their daily work at Amazon as one of the job's most emblematic features: now here, now there, now down, now up, with no hope for rest and none for lesser pain. Pickers commonly use words such as "pulling," "pushing," or "running" to describe the extremely physical nature of their daily routine. The 2020 coronavirus crisis has only made the process more taxing. In an interview with a local newspaper, a warehouse associate also involved in worker advocacy in the Toronto area described the effects of social distancing in a workplace where employment relies on meeting quotas: "Our performance is evaluated based on the number of orders picked, but at the same time we are discouraged from entering the same warehouse aisle to comply with social distancing. If our performance is poor, we risk losing work."[20]

Old-timers who have seen generations of associates come and go know this all too well. Take Peppino. He is tired, and he is angry. A man in his 50s, Peppino has worked in several different sections of the warehouse, but picking is the process he hates the most: "After a while your back breaks down, you develop hernias, carpal tunnel, stress-induced psoriasis. There are 20-year old people in there who look like my 80-year old mom." This may sound overdramatized, and indeed the story comes from a worker who has been embittered by years of work, but Amazon itself knows that injuries are a major problem in its warehouses. Internal corporate reports about safety in US warehouses show an increase in injuries between 2016 to 2019. In the last year included in these reports, Amazon recorded 14,000 serious injuries, with an overall rate of 7.7 serious injuries per 100 employees. This is nearly double the industry average, as confirmed in a 2021 report that blames the company's "obsession with speed."[21] Seasonal peaks only make things worse. According to data released by Amazon, injury rates tend to spike around Prime Day and Cyber Monday.[22] A report by the American National Employment Law project found that the most common injuries involve "sprains, strains, and tears to the shoulder, back, knee, wrist, and foot" that "can stay with workers for the rest of their lives, leading to chronic pain and […] long-term disability."[23] Peppino recalled how management does not necessarily like the stops that come with injuries: "Do you pass out? Feel unwell? Boss, shall we call an ambulance? Nope, they don't like that. Once, a co-worker passed out and was not recovering" and management did not call an ambulance until "her husband arrived, he worked there too, and started breaking chairs and throwing them around, freaked out, *then* they called an ambulance."

In the face of the growing problem of musculoskeletal injuries caused by the repetitive nature of tasks such as picking and packing, Amazon predictably goes for palliatives and technological fixes instead of rethinking the rhythms of work. Workers have even reported that some Amazon FCs sport vending machines selling ibuprofen pills. In his March 2021 letter to shareholders, Jeff Bezos announced that on top of Amazon's WorkingWell program, which coaches small groups of employees on safety and body mechanics, the company is developing "new automated staffing schedules that use sophisticated algorithms to rotate employees among jobs that use different muscle-tendon groups."[24] In sum: another algorithm will decide when you have put too much pressure on your right ankle and move you to a job that will impact your left wrist instead. And the issues go further. The repetitive nature of the tasks assigned by the scanner can add another type of strain, one that no algorithm can prevent. As Peppino put it, once workers acquire the "conditioned reflexes" needed to do their job, the work becomes mindless. "You just need to follow the scanner, which tells you: 'go here, go there, pick this and pick that.' You don't need to do anything else, don't need to think. Eight hours can last 24 hours because you are in a limbo." In effect, Peppino and others like him perceive algorithmic control of their work as something that takes away their autonomy, governing their actions down to the smallest detail. Working in the warehouse can also be a lonely and alienating experience: every day, pickers spend eight or ten hours carrying out hundreds of short trips along the aisles of the pick tower while trying to keep up with the unreasonable speed required by management. The lack of sociality does not help. Workers, especially the precarious, can barely take bathroom breaks, and are discouraged from stopping to make small talk with their co-workers.

Amazon calls its warehouse associates the "heart and soul" of its operations, but because of the standardized, sped-up, and algorith-mically managed nature of the work, many workers see themselves as mere appendices of technology. An Amazon worker from Seattle perfectly encapsulated the resulting dynamic in a comment posted to Glassdoor, a website where people review and rate their employers: "The pickers and stowers like me get treated like faceless cogs in the machine, despite the fact that cogs are what makes [a] clockwork run smooth to begin with." When they rally behind the slogan *We are not robots*, as they have done in recent years, Amazon workers communi-cate the feeling that they aren't only treated like robots, but are actually *becoming* robots—always in motion, ready to obey any order sent through their barcode scanner, and never complaining or stopping for a break.

At MXP5, the rhythms of work have changed over time. On the one hand, targets tend to increase, thus speeding up labor, but on the other hand unionization and worker resistance have the opposite effect. A few years after the first strike in 2017, workers in Piacenza reported that more and more employees limit their adherence to the Amazon pace. Full-timers enjoy legal protection against managerial enforcement of speed, but even some temp workers have slowed down, Peppino told me in 2021: "More and more workers take it easy. They get the job knowing they are only going to be here for a month, and thus don't bother running too much. They know management must be more careful now that there is an established union presence in the warehouse." These are clearly workers who do not bet too much on a contract renewal, let alone on being moved to a full-time contract. But, Peppino continued, "they all know someone who has worked here, they know the drill, and decide it's not worth it." Of course, there are those who defend or enjoy the work even when it is intense. At least initially. In another online comment, an FC associate from Virginia highlighted that "As long as you're able-bodied, and are prepared to bust a sweat, it's a great job with good starting pay. I regularly get 14,000+ steps at work, every day. I call it getting paid to workout." This type of reaction tends to come from young male workers, especially those in their first months of work. For many others, the Amazon pace is less appealing. When I interviewed her, Barbara was in her late 40s and had already developed hernias working at MXP5. She noted that "you'll find the bionic dude who is 20 years old, and at home his mom takes care of everything. For him eight hours and 15 kgs are like going to the gym. Good for him." Barbara started working at MXP5 after losing her job as a designer. Hers is not an uncommon story in the sluggish Italian economy, nor would it be odd in other places where the middle class never fully recovered from the 2008 crisis and the austerity politics that followed. She recounted to me that on her first day, the instructor assigned to her group of new hires "immediately showed us the so-called Amazon pace—that is, that you have to be quick." But her body would not abide by the rhythms demanded by the machinery and by the fine-tuned use of time–motion analysis in the warehouse. "I work a lot with my right hand, and [...] I got two or three tellings-offs because instead of grabbing the item with the left hand I would get it with my right hand, and in doing this I was taking a fraction of a second more," she explained.

Technology does not always make things easier. The instructions given to pickers are designed to be simple and easy to interpret. But lots of problems can get in the way of completing a batch on time and making rate. For the masses of workers who use barcode scanners 24/7, especially during peak seasons, the scanner can quickly move from aid to obstacle. Barbara saw Amazon replace waves of worn-out scanners. But, she said, sometimes they remain in use long past their prime.

> Many of the scanners we use have issues, perhaps a broken screen or a problem with the sensor, so they don't scan codes correctly— maybe they fail to scan three out of ten barcodes. So you keep on repeating an error because the system rejects your pick. And you can't pick, because [the scanner] does not read the barcode.

This, she said, can result in management reassigning you another copy of the item in a different area of the pick tower—so you need to walk there, which takes time and slows you down. Even when the scanners are working, there can be problems with the inventory system itself, all of which affects one's ability to make rate. "They give you an image to facilitate your search, maybe it [shows] a black T-shirt, but actually you are looking for a white T-shirt, so good luck! You unloaded a truck with 1,500 white T-shirts, but the one in the picture is black. This seems trivial, but you get quickly lost in these issues." Barbara also found the scanner itself difficult to operate when things didn't go smoothly. Among the functional peculiarities, many commands are in English. "My son would learn this stuff in seconds," she told me. Then she handed me a cheatsheet she wrote to translate technical terms from English. She had kept it hidden from management by folding it in her badge holder. "D+enter = article is Damaged"; "M+enter = article is Missing"; "Q+enter = Quit," which is what you type in when you want to log out from the system. The list goes on.

A further feature of outbound work illustrates how the warehouse's algorithmic organization of labor breaks down the fulfillment process into simplified, standardized tasks in order to thoroughly control it.

As pickers work on batches generated by the system algorithm, they are not necessarily assembling a single order for a single client. Again, knowledge about the ultimate composition of a particular order resides with system algorithms. As Peppino told me, "anyone can work on any order. You don't get to know what you are working on." When you ordered the coffee mug, eyeliner, and USB drive, the three items may have been picked by three different workers as part of three different batches. Thus, both picking and inventory are chaotic. The task of re-ordering is left to a smaller subset of outbound workers who, themselves guided by algorithmic management, "rebin" batches—that is to say, they re-compose specific individual orders. They receive carts full of totes from pickers and must scan all the items they contain. The system then outputs directions to a computer screen, prompting them to place the items into particular "rebin cells" nested in a yellow shelf. Once your coffee mug, USB drive, and eyeliner have all arrived and been placed in the appropriate cell, the order is ready to be boxed and sent out. Workers do not take care of the whole process, but rather perform individual tasks strictly dictated by algorithms.

THE ROBOTS ARE COMING

Amazon steadily introduces more and more physical automation to complement its managerial algorithms. But when robots enter the picture, they do not necessarily make warehouse work easier. They certainly speed it up, increasing workers' productivity, and also the rate of injuries and feelings of alienation. While many features of labor at Amazon are standardized across the company's global facilities, some warehouses are more automated than others. In 2012, Amazon purchased a start-up called Kiva and renamed it Amazon Robotics. A minority of fulfillment centers are now equipped with the company's roboticized shelving system. In these facilities, workers do not enter aisles to pick, but are instead served by robots that look like the Roomba vacuum cleaner's bigger, more stubborn siblings.

Round, flat, and predictably orange in color, these robots—popularly called "Kivas"—flit across warehouse floors carrying tall yellow shelves fitted with dozens of individual cells. Algorithms organize their movements, sending the robots to fetch the right shelf and bring it directly to the correct stower or picker. Photocells mounted on the robots' undercarriage enable them to follow special paths drawn on the floor. Associates stand at workstations equipped with scanners and computer screens. As shelves arrive, the stationary workers pick and scan the commodities as per instructions received through the computer. In the meantime, other robots are lined up, ready to take the place of their predecessor as soon as the picker is done with it. Due to the speed and size of these robots, special precautions need to be taken. In robot-equipped warehouses, metallic fences divide workers from the area where hundreds of the robots zip around with their shelves. Kiva robots are used in FCs that trade in "sortable" commodities, small items that can be moved by hand. The small cubic cells lining the shelves they carry are designed to store items no larger than a school textbook. Another bigger and newer Amazon Robotics product called Hercules is used in "non-sortable" FCs. These can deal with larger stuff, like television sets and bicycles. Other robots are being introduced, too: the Robostow is used to lift pallets or boxes around. The Pegasus is a more advanced version of the original Kiva.

The idea of moving work to the workers rather than the other way around is certainly not new. The managers of 19th-century American slaughterhouses developed the method, driving live animals into the building to be killed. Their carcasses, too, were then moved through different areas where workers handled various aspects of producing the final meat cuts. Ford later adopted this continuous-flow production process in its factories, this time in the assembly of cars, rather than the disassembly of livestock. In both slaughterhouses and automobile factories, this method increased the efficiency of the production process and broke down the tasks assigned to each individual worker: unlike craftsmen, assembly line workers only need to take care of a single step in an otherwise extremely complex process. The same is true at Amazon: robotized workstations speed up labor and simplify work. They also further dispossess workers. Kiva stations, for instance, mean stowers and pickers no longer even need to know the geography of the pick tower.

Amazon's robots don't only enable a stationary workforce, they also allow the warehouse to store commodities even more efficiently, as the shelves can be squeezed against each other without the need for aisle space: robotized warehouses can hold double the inventory per square meter. But robots cannot obviate human workers. On the contrary, even in partially automated workplaces the labor of storing and retrieving items remains manual and repetitive, and abides by the same chaotic logics that are at play in warehouses not equipped with Kiva robots. In fact, the presence of robots does not necessarily mean that a fulfillment center is newer. Amazon is still building non-robotized facilities in areas where regional commercial and logistical considerations make it financially prudent. The robots simply change the physical nature of the work, as associates are not required to walk for hours in the pick tower.

This was readily noted by Tina and Giorgio, two young associates who had been working at FCO1, a robotized warehouse near Rome, for two years when I spoke with them. I first met them in the city center, in the offices of a union that was organizing the warehouse's associates. They liked how Kiva robots eliminated the need to walk all day. But at the same time, they lamented the similarities between the workstation and a cage—a comparison they weren't alone in noting. "One time the manager passed by and asked me, 'would you like some peanuts?' They even mock you, you know?" relayed Tina. "When you pick, you are in a cage. You are practically a monkey." Giorgio also noted the extremely repetitive nature of the work. "You keep doing the same actions, up and down, up and down," as he put it. And indeed, the workstations do resemble cages, surrounded as they are by grids that isolate workers from each other and protect them from the robots.

Kiva workstations are a common topic of conversation at MXP5. While it is not automated in this way, some MXP5 workers have visited robotized fulfillment centers. In what was an unusual glimmer of optimism, at least for him, Peppino told me that "those of us who have been in Seattle have seen what may be the best FC. Very automated, so physical efforts are reduced by a good 50 or 60%. The worst jobs are still there, but the human is aided by the machine." Many others were more pessimistic. Maria reacted strongly to pictures she saw of Kiva warehouses and stories she heard from FCO1 workers. "We are worried because with these little robots you are standing in a workstation [...] locked in a cage alone." The robot, she continued, "brings to you the shelves, a tablet tells you where to put stuff, so you click, store and pick from this workstation" with no contact with your co-workers. What Maria saw in the robots was an even more alienating future warehouse: "There may be no physical effort, but after a week in this cage with a computer I would be mentally depleted," she concluded.

The alienation Maria described is undoubtedly bad, but the introduction of robots creates more directly material problems also: Robotized warehouses are significantly more dangerous. Because Kiva robots are used to speed up work, workers find themselves trying to catch up with quotas that may be up to four times higher than in a non-robotized fulfillment center. For instance, pickers have reported increasing their rate from 100 to 400 items per hour after robots were introduced. Tina and Giorgio said that at FCO1 they were now expected to pick 500 per hour. As a result, warehouses equipped with Kiva robots have injury rates 50% higher than those without robots.[25]

APPETITE FOR DATA

The use of technology does not allow Amazon to do away with its workers. If anything, it increases capital's need for the "real people and real bodies" that labor in the warehouse, as put by Ursula Huws.[26] When Marx said that machines create "new incentives" that whet capital's "insatiable appetite for the labour of others,"[27] he had in mind the factories of 19th-century industrial capitalism. But the same appetite remains in today's digital capitalism. Thanks to its ability to use software systems to control and direct workers, Amazon mobilizes masses of workers who are quickly put to work in the warehouse, where they perform repetitive physical tasks for a few weeks or months, and can lose their job just as quickly. But Amazon is different from 19th-century industrial capitalism as well. The increasing preponderance of digital mechanisms that feed off workers means that the warehouse craves not only the labor but also *the data of others*. The software systems that underpin inventory and fulfillment need to continuously turn everything workers do into data. In turn, the information generated is crunched by distant algorithms and subsequently used to break down, reorganize, and strictly control the labor process. In Amazon's digital factory, technologies change, but the fundamental dynamic of monitoring labor in service of efficiency, control, and the interchangeability of individual workers does not.

This capitalist strategy is nothing new, and has been remarked upon time and again by those studying labor in industrial capitalism. For example, in his 1970s study of factory work, sociologist Harry Braverman described management's ability to gather knowledge and use the monopoly over such knowledge to "control each step of the labor process and its mode of execution."[28] Braverman saw automation as a tool to perpetuate and consolidate the authoritarian structure of the factory. Analyzing the labor process at FIAT in the 1960s, workerists also pointed out the extent to which the assembly line, by allowing tasks to be broken down and standardized so that each worker only has to perform a single repetitive action, functioned in service of capital's domination over labor.[29]

In contemporary digital capitalism, widespread data collection is certainly not unique to the Amazon warehouse. American scholar Shoshana Zuboff has described an emerging form of "surveillance capitalism" based on the pervasive collection of data from all users.[30] Zuboff gives a name to what we now all know very well: that if we own a smartphone, everything we do as we walk around, shop, or even chat in our living room is recorded, turned into digital data, crunched by algorithms, and used to sell us stuff or to control us. Others, like media theorists Nick Couldry and Ulises Mejias, have proposed that this amounts to a new form of colonial relationship: data are "extracted" like natural resources to the benefits of a handful of corporations.[31] Everything we do—Zuboff calls it "behavioral surplus"—can generate value for digital companies like Google and Facebook that engage in the widespread capture of user data. Amazon itself demonstrates how customer data are harvested by tech companies. As people shop through Amazon's e-commerce websites or at Amazon Go automated grocery stores, talk to their Alexa or watch a show on Prime Video, their activities are recorded, analyzed, and used to improve the service or encourage further consumption. Almost any behavior, even the most incidental ones, can be datafied and valorized by digital platforms.[32] It is just there for corporations to take.

In the workplace, it is not so simple. Workers are not just the objects of surveillance: they must be pushed to perform the specific processes that generate valuable data to be incorporated into the machinery, as the labor process is integrated with and run by data collection systems that would not function without their labor. In sum, workers have agency. In the warehouse, data collection can't take place without pickers' and stowers' sustained physical work. The workerists saw this dynamic unfold in factory work, and presented the relation between labor and capital as one where capital struggled to control the source of its value, that is: workers. As put in the early 1960s by political theorist Mario Tronti, "a history of the industry cannot be conceived as anything other than a history of the capital organization of productive labour" since "it is productive labour which produces capital." It follows, he argues, that "the capitalist class is, from its birth, subordinate to the working class. Hence the necessity of exploitation."[33] In sum, capital depends on labor and needs to control workers and secure collaboration in production processes in the face of their unruliness. It cannot afford workers who resist the rules and processes set by machines, let alone workers who slow down or strike. In industrial capitalism, that meant that employers had to convince workers to follow the rhythms dictated by the assembly line. Digital capitalism must persuade workers to obey the commands of an algorithm, and follow a dictated pace. In both cases, it is about forcing workers to synchronize and calibrate with machines.[34] To work hard.

The generation of data from workers' activities and its incorporation in software systems does not eliminate the need to control labor.[35] If anything, the need to more rigidly control a workforce grows with the increased technological nature of the labor process. At its core, this is a purely financial factor: from capital's viewpoint, the more expensive a machine is, the more efficiently it has to be used, or its value will be wasted. The more Amazon relies on automation and algorithmically-captured inventory, the more it needs to find ways to push its workers to become robots that seamlessly and efficiently abide by the rhythms imposed by automation. What if they refused? To minimize this risk, Amazon deploys a unique set of managerial techniques. Datafication allows management not only to eliminate the need for workers' knowledge of inventory, but also to surveil them, monitor their productivity, and sometimes discipline them. Amazon has innovated the use of technology in the workplace, but still uses it to prop up forms of despotism reminiscent of the tumultuous days of early industrial capitalism. At the same time, its management strives to create a workplace culture that ensures workers' collaboration with the goals of fulfillment.

3

Have fun

The barcode scanner is not just the main tool used to organize warehouse work. It is also at the center of the managerial techniques Amazon uses to monitor employees and persuade them to align their work with the warehouse's fast rhythms. Upon logging into the system, workers make themselves completely visible to management, subject to a pervasive system of control.[1] The barcode scanner that assigns them tasks ("pick the stuffed bear in cell X, aisle Y") is also used to surveil their every movement, recording information supervisors can use to see how quickly they work and how often they went to the bathroom. This information might lead to a message popping up on the screen of a worker's scanner: "Meet team lead for a feedback session," as managers and supervisors (called "leads" at Amazon) act on the data to discipline and fire. In other instances, the barcode gun delivers questions and polls: "How do you feel about working at Amazon?" The scanner functions as a harbinger of ideological, not only physical, control.

The techniques enabled by the scanner, together with other managerial tools used by Amazon to monitor and persuade its workers, to manage their bodies and minds, are key to running the warehouse. The truth is, we tend to see workers as depending on their employers, but capital's need for labor is much higher. The blend of control and consent used by Amazon to fulfill this need is, of course, a common feature in capitalist relations of production, as employers must counter workers' tendency to resist the commands they are dictated to. After all, workers have interests that are different from those of capital, whose bottom line is profiting from their labor. But at Amazon, management relies on a unique mix of brutal, conspicuous disciplinary action and more subtle attempts to motivate worker self-discipline through play, psychological nudging, and the promise of happiness. In this way, workers in the warehouse are simultaneously managed by fun and by stress.

The company is famous for its uniquely pervasive system of digital surveillance. The barcode scanner is one of the other components of this system. Workers are tracked through widespread cameras, body scanners, and the monitoring of the social media pages they use to organize. All of which helps the company create a climate of despotism mediated by technology but enforced by warehouse supervisors and managers. Ultimately, it is these supervisors who are in charge of scolding workers if they are not fast enough, enacting aggressive anti-union tactics, and disciplining employees who are subject to a constant threat of dismissal or non-renewal of contract. But those are not their only duties. In fact, focusing strictly on punitive measures fails to capture the truly radical nature of Amazon's managerial culture. While fear and stress are certainly part of the equation, the other elements are informality and a culture of staged fun. Managers and supervisors are tasked with turning the warehouse into an engaging and informal workplace: Amazon draws on human resource techniques designed to more subtly direct worker participation and engagement to the benefit of corporations. This means Amazon workers spend their lunch breaks in canteens furnished with colorful sofas, foosball tables, and arcade games. They are prompted to sing and stretch at daily briefings. They may wear garlands on Hawaiian day. The computer screen in their workstation might sometimes ask them to "project loving energy" as they serve the robotic pod that brings them products to pick.

This artificial informality seems to clash with the harsh reality of other aspects of warehouse work. How do you reconcile the pervasive surveillance and sometimes open despotism of the dark aisles of the pick tower with the bright and colorful break rooms where management strives to present the warehouse as a special and fun workplace? This contradiction generates dissonance between worker expectations and the reality of the job. As one former warehouse worker from New Jersey put it in an online comment, "The presentations [and] the first-day orientation were too good to be true. A utopian worker's paradise was promised by the hiring team and the management staff." Yet, the commenter continued, they were ultimately "employed as a warehouse picker and it was a boring, mind-numbing, and thankless task." In fact, the imperative to have fun, as the company workplace slogan goes, is suspended when necessary. It's a stick and carrot system, really, deployed to make sure that workers keep up with the productivity levels, flexibility and physicality of the job. To control them. If we look at the warehouse in this way, the coexistence of a pervasive surveillance and managerial domination on the one hand, and a culture of staged fun and participation on the other, is only apparently a contrast. Workers may experience it as odd and paradoxical. But for the firm, there is no contradiction. From Amazon's perspective, both fun and stress are effective.

THE WAREHOUSE AS A PLAYGROUND

"One. Two. Three: outbound!" At the beginning of each shift and conclusion of every lunch break, all workers are required to attend a management-led "briefing" (or "standup," in the US). These mandatory five-minute rituals are used to impose participatory practices into Amazon's workplace culture. For example, workers might be asked to raise their hand and suggest a "success story" in front of the rest of the team. Having one ready can be rewarded with a round of applause. Workers are expected to cheer, sing, or even dance. Managers often comment on team performance, and workers are implicitly required to celebrate.

Zak had only worked at MXP5 during one seasonal peak, but still remembered motivational briefings where managers would say things like, "Yesterday we had an insane productivity rate!" followed by applause. Another time, he told me, "A manager had come back from an experience in a different warehouse, and we applauded too." This exercise came to prominence during the coronavirus pandemic in 2020, as briefings were used by supervisors to motivate and reassure workers concerned about virus outbreaks. In fact, the briefing is perhaps the most visible element of Amazon's attempt at building a culture of joyful informality in the warehouse. It is a hallmark of the company's top-down imposition of a workplace culture that celebrates warehouse work, and brands it as cool and participatory. One must be happy when working at Amazon.

Some workers like these briefings. Elisa, a young temp worker I met, was initially puzzled by the practice, finding it odd. But she described a change of attitude a few months into her job at MXP5, as she found the briefing to be the only space where she could find a collective dimension to contrast the solitude of the pick tower. "Slowly one gets used to it, and psychologically it is useful because, in fact, in there you see [...] so many people, people from diverse ethnic backgrounds and different walks of life, and yet it's all smiles. They [Amazon] create it, they create this setting, and it works." To illustrate, she gave me the example of managers knowing and using workers' nicknames—the same artificial closeness we experience when a Starbuck barista calls us by our chosen name when our soy latte is ready. In my FC visits I witnessed other mechanisms that contribute to this studied informality. For instance, at MXP5, only mechanics, truck drivers, and other blue-collar-type laborers were referred to as "workers." The mass of pickers and stowers were "boys" and "girls."

Yet, while some workers appreciate these practices, other associates are disillusioned about the briefings and other activities aimed at promoting warehouse culture, especially when they are clearly linked to the need for productivity. For instance, it is not uncommon to hear workers ironically characterize the briefings as "dog and pony shows" or "Alcoholics Anonymous meetings" as they describe cracks in the warehouse's culture of fun. Not everyone can afford to be so cynical, even if they would want to be. Full-time employees enjoy more labor protections than temp workers whose contract renewal may depend on their productivity and ability to show adherence to the culture of fun. Nevertheless, the workers I spoke with expressed a range of misgivings. Emma, a temp picker in her 50s, described how workers are asked to exercise and stretch to prep for the physical work they will sustain in the pick tower. Often gender plays a role in who is chosen to lead the exercises. Emma said that "they always call young girls to stretch" in front of everyone. After all, "leads and managers are almost all male," she remarked.

Rank-and-file workers often expressed animosity when telling me about the briefings. Luca, a full-time outbound worker, said that the briefing "is motivational: they tell you, 'So far we only picked so many pieces, so in the last part of the shift you need to do more.'" He became animated as he shared his reaction: "They try to incite you: 'Guys, we have 200,000 pieces today, are we going to make it?' All bullshit that is only there to make you run more, to exalt you." He was clearly fed up with the briefings he had been exposed to twice per shift for the three years he had spent at MXP5. He observed that management never tells workers when a target has actually been reached and the team can finally slow down.

Workers often encounter this paternalistic culture even before walking into the warehouse, as the job is advertised and pitched in moral terms from the get-go. At a recruitment event in Toronto, the Amazon rep told prospective workers that Amazon work is all about customers and their needs. The recruiter illustrated this by telling a story: a consumer had ordered a Christmas present for their child, and for some reason it wasn't delivered in the promised timeframe. But Christmas day was rapidly approaching. Something had to be done. And so, the rep relayed, one employee personally drove the package from the warehouse to their house. "True story [...] happened some-where in the States," they concluded. Some workers describe their reaction to these moralizing attempts in unambiguous terms. "Manage-ment may round you up and tell you things like, 'Think about it folks, thanks to you today many children will smile, you have brought joy in the homes of thousands of families.' That's when I want to headbutt them," explained an embittered Luca. Perhaps Luca could change his attitude by participating in AmaZen, the meditation program intro-duced in US warehouses in 2021, which "guides employees through mindfulness practices in individual interactive kiosks at buildings [...] including guided meditations, positive affirmations, calming scenes with sounds, and more," as per corporate press release.[2] Zen or not, ambivalence or even cynicism in the face of this normative pressure to be positive is not uncommon among workers. Cynicism, in particular, can be a form of disidentification that enables workers to reclaim agency over the construction of their identities. It also comes easy. There is, after all, a glaring dissonance between the promise of a happy workplace and the daily reality of FC work. As one disillusioned worker from Indiana described in an online comment, there is "not enough time on breaks to really bother with the video games in the one break room."

There are no arcade games at MXP5 in Piacenza, but as in all Amazon fulfillment centers, its break room reinforces the playful and informal ambitions of the warehouse. Sofas in basic reds, yellows, and greens sprawl beside foosball, table tennis, and a big-screen TV. Flyers advertising the next group activity or team pizza night line the walls. As I walked through the canteen of another fulfillment center,

I couldn't help but think of the Googleplex in Mountain View, the colorful campus where Google engineers work surrounded by giant dinosaur skeletons or replicas of spaceships, and dine in fancy cafeterias. MXP5 is punctuated by special days designed to boost the feeling of community. Hawaiian day, chocolate day, 1990s day—the list goes on. This is not unique. At PDX9, a warehouse near Portland, a dedicated team prepares balloon art to be displayed in the FC, and orange Kiva robots hang from the ceiling. Like the Googleplex, the warehouse is built to look like a playground: an informal and fun workplace. The image of the tech start-up loaded with toys, bean bag chairs, noise-cancelling nap rooms, and graffiti murals has become a cultural trope—a symbol of the entrepreneurial spirit of technological innovation and self-made fortunes.[3] *Stay foolish*, to use Steve Jobs' famous motto. Amazon is anxious to tap into this cultural appeal, and nowhere is it more apparent than in the architecture of warehouse canteens and break rooms.

And it's not just about the furniture. On my first visit to MXP5, I was surprised to hear loud music playing in one of the sorting areas. Meanwhile, the worker serving as an ambassador and showing us around that day stressed the laxity of Amazon's dress code. Piercings, shorts, and colored hair are all welcome here, she noted—a policy that started in Amazon's very first warehouse, in Seattle. This dress code may not be that different from many other workplaces. But taking a guided tour in an FC or reading worker testimonials online, it is difficult not to notice the frequency with which the dress code is mentioned. Amazon seems to place importance on it as a symbol of the company's benevolence. At times, the dress code becomes an even more pointed fixture of Amazon's have fun ethos. An online comment from a worker described how new employees noted that during their orientation, a big deal was made of the fact that on special days they could come to work wearing onesies or pajamas. The commenter recounts how the worker leading their orientation told them: "'I didn't know I had a boss that owned a Spongebob onesie,'" observing that "the guide jokes in a tone that suggests she's delivered that line a hundred times before."[4] Workers must also learn to speak "Amazonian," a language made up of corporate slogans like, *Deliver results*. The lingo is so ingrained in FC culture that many bring it home. Just a few months into her MXP5 job, a worker told me that she no longer cleaned her apartment—she now "does some area readiness," as per Amazonian lingo.

The fun attitude, the lingo, the continuous motivational briefings, the break room architecture, and all the other tools used to ensure the adhesion of workers to corporate culture: none of it was invented by Amazon. Employee happiness is the topic of an entire genre of management literature, and daily briefings have become a fixture of several big corporations that rely on team-based work. Maybe not in other workplaces in Piacenza, but similar briefings are common even in advanced manufacturing, for instance, in FIAT auto plants in contemporary Italy. In many ways, Amazon follows a playbook advocated by management gurus who encourage workplaces to foster a *work hard/ play hard* environment. Informal dress codes, office parties, games, and humor—all of these are advanced as hallmarks of good management, supposed to increase worker motivation and creativity, while simultaneously countering anti-management sentiments and stress. Organization studies theorist Peter Fleming calls it a "managed culture of fun," with the ultimate aim of portraying even menial work as a calling.[5] All this is to say, Amazon's culture is not emergent from social interactions, but is designed from the top down—team leads and managers are tasked with continuously enforcing it.

Nevertheless, some workers resist these overtures and find ways to work around the culture of managed fun to build their own spaces of socialization—be it in the break room, in quick meetings in the aisles of the pick tower, or outside the warehouse. It is not always easy. Indeed, while workers can (outside of their shift) relax and socialize in the canteen and other common spaces, many describe a work experience in which "human interactions are discouraged and disincentivized, if not explicitly punished," as seasonal MXP5 picker Zak told me. "In the end, even in the canteen there is always a lead or a manager."

Efforts to align workers with corporate culture often begin in the hiring process itself.[6] Elisa still remembered the test she was administered by the temp agency that hired her:

> Even just the aptitude test is composed of peculiar questions: "Do you feel positive? Are you luckier than other people? Do you feel happy every day?" This is peculiar; what do you answer to the question, "Do you feel ready?" Ready for what? And yet you must

answer because you know in the first place that they want a positive person who feels lucky and ready.

While personality tests are common aspects of hiring in several industries, at Amazon this sort of testing does not stop at recruitment. It continues in the warehouse, where daily briefings are supplemented by a continuous stream of questions asked through the barcode scanner.

What is tested is the worker's compliance with Amazon's system of mandatory fun—thus nudging the worker into alignment. But Amazon frames the tests instead as tools for worker empowerment. Amazon describes this "Connections" program, as it is called, as a "real-time, company-wide employee feedback mechanism designed to listen to and learn from employees at scale to improve the employee experience. Each day Connections questions are delivered to every Amazon employee on a computer, a workstation device, or a hand scanner."[7] In 2020, Amazon claimed to receive over half a million responses daily, in 21 languages, from employees in more than 50 countries. Connections, the company continues, "analyzes response data and provides insights to managers and leaders to review and take actions as they uncover issues or see opportunities to improve."

The following example, from a story posted by an American associate on a blog in 2017, illustrates how the Connections system works:

[...] my shift has just begun. I carry around a small device called a scanner. It asks me:

How do you feel about working at Amazon?

(1) Great!

(2) Great! I'm proud to work at Amazon!

I'm about to select (2) as always, but I recently figured out that there's a way to scroll down the screen (it involves an orange button and the number 8, in case you were curious). This reveals two more answers:

(3) I wish I was working a job using different skills.

(4) Prefer not to answer.

"Huh, that's funny," I think to myself. "I wonder how many other people never realized that there were extra answers." At any rate, the correct answer hasn't changed, and so I select it [...]

My scanner has another question for me:

How do you feel about this statement?

"Amazon gives me all the training I need to do my job successfully."

(A) strongly agree

(B) agree

(C) neither agree nor disagree

I'm about to respond, but mindful of my previous experience, I realize that there might be additional hidden answers. I scroll down and, what do you know, there are two more options:

(D) disagree

(E) strongly disagree

As usual, the correct answer was in the first set. I select it and continue with my shift.

To interrupt the routine, I go to the only place in the warehouse without cameras in plain view: the bathroom. Inside, a new factoid is posted above the urinal. It reads:

When asked whether they had all the tools necessary to do their job correctly, 82% agreed or strongly agreed! If you ever feel you do not have adequate training, please contact HR.

I'm stunned. 18% of the people did not give a positive response to an obviously loaded question that might threaten their company prospects? You respond to the question *after* logging in, so it's not like they don't know who you are.[8]

Workers report that they experience these questionnaires as a form of ideological control. Since they do not trust that what they answer will be kept anonymous, workers fear that the tests can be used by management to discipline them individually, even though Amazon states that "employees may choose to answer or not answer any question and individual responses are aggregated and shared with managers at the team level to maintain confidentiality."[9] Nor do employees trust other top-down managerial practices that aim to involve workers in a caricature of workplace democracy and participatory decision-making. For instance, management provides whiteboards called "Voice of the Associate" boards, which are designed for "providing employees a forum for expressing their concerns, offering suggestions, and asking questions on a daily basis to leadership. Leadership teams reply directly to questions, promoting dialogue and efficient remediation of issues."[10]

During one of my tours, the guide described how the company had in the past changed work processes or solved problems based on information collected through the boards. A famous example, we were told, is the placement of boxes for storing inventory on bottom shelves. Workers can pull them out like drawers—enabling them to store or retrieve a commodity without getting on their knees.

Workers often flip this picture, though: many pointed out that negative comments are largely ignored. At MXP5 and in other fulfillment centers, many have told me about times when assertions like "We need a union," and questions like "Why did we lose pandemic pay?" or "why are we called essential workers but you see us as disposable?" were posted to a Voice of the Associate board—only to disappear immediately. Apparently, management did not consider those ideas innovative enough to be entertained. But Amazon seems to be innovating their approach to employee feedback itself. In many FCs, including in Piacenza, the whiteboards have been taken away and replaced with computer screens: workers now need to be logged into the system to leave a comment, so that management can decide which messages are shown and identify the workers leaving comments. Problem solved.

ALL THIS FOR A KEYRING

No playground would be complete without games—not even a playground designed to ensure a smooth, productive, and ultimately fast mode of relations between workers and machinery. Thus it is no surprise that gamification features as a prominent element of Amazon's culture of mandatory fun. Gamification is the process of merging enjoyable aspects of game playing with productive activities. Or, as one definition from management theory describes it, the use of an "employer-imposed game in a work environment where the goals of the game are designed to reinforce the goals and purpose of the employer."[11] It goes beyond the presence of foosball tables or arcade consoles in the break room, and aims to directly influence work itself. Masquerading as a form of play between workers, it not only adds to the culture of mandatory fun—but also introduces competitive elements that may have an accelerant effect. Gamification in the warehouse takes many forms. The barcode gun itself is laden with game elements: pickers must speed up to catch the next blip on the screen, to pick the next commodity before the bar on their screen dries up—90 seconds, 60 seconds, 45 seconds, will you make the target? But in a growing number of American fulfillment centers, Amazon even uses actual video games to put employees into playful competition with one another. The computer system translates workers' physical actions, such as picking items, into in-game, virtual moves. So, the faster someone picks items and places them in a tote, for example, the faster their car will move around a virtual track. These simple games, which have names like PicksInSpace, Mission Racer, or CastleCrafter, are displayed to the workers on tablets positioned in the workstations where they store or pick items from Kiva robots. In some cases, workers win virtual "swag bucks" which they can exchange for material rewards, like Amazon-branded merchandise.

These are not the only games played in the service of higher productivity goals. Managers are also able to launch "power hours"—hour-long competitions in which all workers on a team are required to "pull," or work as fast as they can. Meaning: pick even more items per hour than during their already hustled regular shifts. Power hours squeeze high rates from associates, who in exchange may again receive petty prizes, such as branded keyrings or movie tickets, in addition to public recognition in front of the team—earning the usual round of tepid applause.

Tina and Giorgio, the two workers from FCO1 near Rome, explained to me how "maybe they tell you, tomorrow both the morning shift and the night shift will do a power hour, and then they decide what the best team is. And if you win they give you Amazon swag, a water bottle, a T-shirt." Prizes don't go to the entire team but, as Tina put it, "to the person who did more pieces at the assembly line." Giorgio found it "cool" that the prizes were individual, to which Tina replied: "yeah, but don't worry, there is always a power hour. Sooner or later your turn will come." Indeed during peak season, which requires sustained productivity efforts, power hours become more frequent. And supervisors become more nervous about power hours, as they turn from motivational exercise into key moments to keep up with floods of orders. Nevertheless, many workers aren't impressed: full-time employees often take it easy during power hours or even slow down on purpose. It is the temp workers whose contract renewal is constantly pending who must pretend to participate and have fun. After all, they are so focused on meeting their targets that "they don't even use the bathroom," Tina said. "All this for a keyring," she sighed.

These practices are in place across Amazon's global network of warehouses. As one Illinois worker put it in an online comment, "Power Hour rewards often are *very* poor, with first place being $10. Less than you make in an hour to exhaust yourself. And it's in the form of a vending machine voucher. And they probably forgot you won or announced it a week late, and you're getting nothing. Congratulations." Similar techniques have long been common in the service economy, serving as a fixture in call centers. But they've been quickly adopted by digital capitalism: among many others, gig economy company Lyft organizes weekly "Power Driver Bonus" challenges that require drivers to complete a set number of regular rides.[12]

Workplace games are not a new phenomenon. While some gamification techniques described above are explicitly framed as play, more subtle forms have long been encouraged by management. One of the most famous examples comes from sociologist Michael Burawoy's 1970s book on American industrial labor. Workers in the factory he analyzed engaged in the process of "making out," that is, cutting corners and hacking the use of machine tools to speed up their production rates. By turning the labor process into a game and competing against each other or against themselves, workers employed through a piece-rate system managed to improve their output and thus make more money. Michael Burawoy described how this practice of making out first emerged from the workers themselves, as they modulated their control of the labor process in their own interest, for example, speeding up tasks so they could afford to take a break. But ultimately, making out generated more value for the employer than the workers themselves—and in the process also served to channel workers' discontent away from management and toward competitiveness with their peers. In doing so, it diminished the possibility of collective opposition against the piece-rate system itself.[13]

The more recent top-down gamification techniques introduced by human resource management theory serve companies' goals in even more obvious ways. Gamification is a "soft" form of control that pushes workers to increase their pace of work in particular moments and particular ways expressly dictated by management. Gig economy apps such as Uber or Foodora famously incorporate technical features borrowed from gambling. Such technologies do more than just make work "fun." In her ethnography of slot machine players in Las Vegas, Natasha Dow Schüll demonstrated how the implementation of reward schedules in gambling technology does not simply exploit gamblers' pre-existing predispositions—it is actually designed to create, cultivate, and amplify gambling addiction. For instance, slot machines can legally produce artificial results, maybe showing a number of cherries in the lines above and below the one that matters, so that the player is made to believe that they were close to winning, and is thus incentivized to throw in yet one more token.[14] Forms of algorithmic management based on such techniques also incentivize specific behaviors, for example, by strategically increasing pricing to convince Uber drivers to converge on a certain neighborhood.[15] Gamification facilitates capital's need to control the rhythms of work and serves to counter workers' natural tendencies to slow down and take it easy. This is something that concerned Frederick Taylor himself, and modern gamification extends Taylor's techniques—not only proving effective at wearing down worker resistance, but going further by actually motivating workers and speeding up their labor.

So, Amazon gamifies the work experience in an effort to enliven it, make it more fun, and squeeze out every ounce of energy it can from its workers' tired bodies. It tries very hard to make sure this gamification and the control it enables do not feel authoritarian. Still, warehouse games are quite different from the spontaneous and subversive types of play that can appear in the workplace—the games people play because they *want to*, not because they must.[16] Philosopher Byung-Chulo Han wrote about the friendliness of contemporary power: a form of domination that calls forth and exploits positive emotions, presenting itself as freedom. In this new form of domination, Han continued, managers strive to resemble motivation coaches capable of connecting with workers at an emotional level.[17] The warehouse is in many ways a case study in the manifestation of just such a friendly power. But Amazon's culture of managed fun is only one element in a much more complex managerial regime. The power relationships experienced by warehouse workers does not always present as friendly.

YOU ARE BEING WATCHED

In the warehouse, workers are subject to a system of total surveillance. The company deploys one of the most intrusive and sophisticated systems of employee monitoring the world has ever seen. The authors of a recent policy report about Amazon left no ambiguity in their choice of title: *Eyes everywhere*.[18] And yet an online, caps-locked comment left by an American picker from Virginia managed to be even more blunt: "EVERY SINGLE THING YOU DO, YOU ARE BEING WATCHED!" Indeed, the ubiquitous barcode scanners carried by workers register every activity they perform, keeping track of their output and breaks, and making the information readily available to managers and supervisors. And there are other forms of surveillance, too. Hundreds of ubiquitous security cameras record footage from the entire building, and supervisors and spies are tasked with spotting collective organizing and union activity. All of this data can be used to discipline. Spending too much "time off task" ("TOT"), such as taking bathroom breaks, can lead to write-ups and even terminations. In many fulfillment centers, political organizing has repeatedly led to firings.[19]

In his 1975 book *Discipline and punish*, French philosopher Michel Foucault used the concept of a panopticon to symbolize the new techniques of control emerging with modernity. Originally conceived of as a type of prison, the panopticon was invented in the late 18th century by philosopher and social reformist Jeremy Bentham. In his design, the panopticon was a round prison with the cells built around a central watchtower. This architecture allowed a single guard to monitor all of the cells from the tower, without being seen by the imprisoned. Moreover, prisoners could not tell if and when they were being watched. In Foucault's words, the prisoner of a panopticon could only assume they were being observed in this asymmetrical system of surveillance: "He is seen, but he does not see; he is an object of information, never a subject in communication. As a consequence, the inmate polices himself for fear of punishment." For Foucault, modernity expanded the panopticon into daily life, as surveillance regimes and institutional forms of discipline were increasingly applied to the general population. In this sense, Amazon is developing and deploying digital technology that extends and deepens Bentham's panopticon further: in the warehouse, but also in the cars and vans that deliver Amazon packages, workers are constantly watched, recorded, their labor measured, and their activities monitored.

Of course, pervasive surveillance is not unique to Amazon. Castel San Giovanni, where MXP5 is located, is a quiet small town where nothing ever happens, and yet hundreds of security cameras surveil people as they walk around, shop, drive, and even when they go to school—in 2018, the city spent 50,000 euros on surveillance cameras that monitor the entrance of educational buildings, from daycares to the local high school. So in a sense, the warehouse is just another place where workers encounter ubiquitous digital surveillance. Only, their relationship with surveillance technology is even more entangled, as workers can't perform their jobs without technologies like the barcode scanner, which renders them dependent on the very tools that monitor them.

Amazon's monitoring of workers starts at the warehouse doors. Employees must leave all their personal belongings outside as they begin their shift. In most cases, all they can take inside the pick tower is a water bottle. When they leave the warehouse floor, even for a lunch break, they are screened through full-body scanners to ensure they are not stealing any of the commodities they handle. Starting when they log into their scanners—or other digital tools like the tablets or computers used in some workstations—they are surveilled by software systems used by managers to control the labor process. Among the most important is the Associate Development and Performance Tracker (ADAPT), a software that tracks worker productivity and determines how quickly they perform assigned tasks, such as locating, scanning, or packing. ADAPT tracks whether workers are meeting their quotas—the number of tasks they are supposed to perform per hour. Quotas are just one example of the KPIs, or key performance indicators, that are ubiquitous throughout the warehouse and in the logistics industry more in general. The result of these methods to quantify and improve the output of a worker, team, or process is the intense micromanagement of labor.

Passing a certain threshold of time off task, for instance, generates "TOT points." Workers who accumulate too many are subject to warnings and, especially for temp workers, their contract renewal may be jeopardized. Standardized personal rates ignore workers' different needs. For instance, pregnant women are disproportionately impacted by a system that counts bathroom breaks against TOT.[20] Workers criticized the paltry break time—for example, a 30-minute period during a physically grueling eight-hour shift, including the time it takes to cross the massive warehouse to reach the break room. While in most cases management is responsible for deciding what temp worker will have their contract renewed or who will be let go, especially in countries with low labor protections, workers in the United States have witnessed terminations that are fully outsourced to the software system. Imagine finding out you have been fired through a message sent automatically to your barcode scanner. Workers report that this has resulted in wrongful dismissals. Surveillance doesn't only result in top-down disciplinary action. Workers in the US have also stated that management sometimes posts personalized TOT scores for the entire warehouse to see, thus singling out workers and creating peer pressure to perform faster.

The COVID-19 pandemic has pushed this surveillance further into workers bodies, as Amazon deployed technology to curb the spread of the virus to its associates. Weeks after the beginning of the crisis, US fulfillment centers started covertly using thermal cameras to scan employees for fever.[21] Additionally, an AI-powered camera system that Amazon refers to as Distance Assistant was deployed to enforce social distancing. It analyzes the position of workers who walk past it, and if the worker is not maintaining proper distancing from co-workers, they appear on a public monitor with their image circled in flashing red, signaling that they must move away from others. Another camera system called Panorama, which Amazon sells to companies such as Cargill and Fender, automates the surveillance of other COVID-19 related infractions, for instance, catching those who are not wearing face masks. The computer vision models that these systems use could be trained to monitor video feeds for any "unusual" activities.[22] Amazon maintains it is working with machine learning technologists to improve these systems, taking advantage of a health crisis to pioneer techniques that could one day be used to further augment managerial power.

Amazon's surveillance system is not confined to the warehouse. If anything, the warehouse is the lab where new surveillance technologies are introduced and tested before being deployed on other workers, in a race to Amazonize more and more workplaces. The misclassified "independent contractors" who deliver for the company through its "gig economy" app Amazon Flex, for instance, are tracked with navigation software that monitors the routes they take. Their productivity, like how much time they spend on each delivery, is also measured. And, of course, they are also subject to gamification, with the app sometimes pitting drivers against one another. In 2021, delivery workers were asked to install another AI-powered camera called Driveri on their car's rear mirror. This camera turns on as soon as they turn on the engine, and continuously records both the road in front of them and the interior of the car. After acquiring the organic grocery store giant Whole Foods, Amazon expanded surveillance to supermarket workers. An interactive "heat map" assigns each store a unionization risk score based on criteria such as worker ethnicity and turnover rate.[23]

Even Amazon customers are surveilled—not only as they purchase items online, but also inside their homes: Alexa, the AI assistant, listens to their most private conversations. Amazon Halo is an application linked to a wristband that incorporates sensors that track things like the user's temperature and heart rate. In turn, this data is used to provide information about one's wellness—if you buy into the idea that wellness can be quantified and calculated in such a way. Consumers can also purchase surveillance in the form of Amazon gadgets to control their homes or neighborhoods. For instance, the "smart" doorbell called Amazon Ring incorporates a camera, offering consumers the promise of security by monitoring what happens outside their doors. But it also allows Amazon to extend its surveillance regime outside the home with their complicity—generating data Amazon can then offer to other institutions as a service or product. Ever relentless, the company has further ambitions in consumer-directed surveillance systems. In a patent for what it defines as "surveillance as a service," Amazon plans to develop a fleet of drones that monitor paying customers' homes for break-ins and theft.[24]

Private citizens are not the only customers of these surveillance systems. Amazon sells its services to American law enforcement and immigration agencies. Ring doorbells connect to Neighbors, an app that produces a heat map of crime. Hundreds of police departments in the US now use this distributed surveillance system. The company also sells AI-powered facial "Rekognition" technology to US law enforcement agencies. Unsurprisingly, the technology has been found to harbor significant bias. The American Civil Liberties Union (ACLU) found that false matches from Rekognition disproportionately affect people of color, and demanded that "Amazon must fully commit to a blanket moratorium on law enforcement use of face recognition [...] They should also commit to stop selling surveillance systems like Ring that fuel the over-policing of communities of color."[25] Beyond developing direct surveillance products, Amazon supports existing surveillance regimes through its other services. For instance, it provides the web infrastructure that hosts the databases used by the Immigration and Customs Enforcement (ICE) Agency to organize the detention and deportation of immigrants. Amazon engineers have protested this collaboration under the slogan *No tech for ICE*. Artist and scholar (and former Amazon FC associate) Hiba Ali critiqued products like Ring as catering to a "market of purchased safety" where "class intersects with race to reproduce a gated white 'secure' suburbia."[26] And indeed, the history of technological surveillance in general is strictly linked to the desire to control and repress non-white populations. Race, and blackness in particular, are major factors in the ways in which "surveillance is practiced, narrated and enacted," as put by theorist Simone Browne.[27] Back in the warehouse, race is also a major critical factor in the deployment of workplace surveillance. In many countries, Amazon employs a workforce predominantly composed of racialized minorities. This means applying to Black and indigenous workers an extreme version of a system of surveillance that has historically oppressed and traumatized them—and keeps doing so.

Warehouse workers are monitored not only to ensure they keep up with the increasingly unreasonable rhythms required by their jobs, but also for the sake of political control. Job ads posted to the corporate hiring website www.amazon.jobs in 2020 advertised positions for analysts tasked with gathering intelligence on "labour organizing threats against the company."[28] The posts made clear a desire for candidates with prior military or law enforcement experience. Fulfillment managers, too, are trained to watch for labor organizing. A video leaked in 2019 demonstrated how Amazon instructs its supervisors to spot early signs of labor organizing, such as workers mentioning a "living wage." Amazon's deployment of professionals trained in the techniques of authority is well established. In late 2020, *VICE* broke a story about Amazon's Global Security Operations Center, a department staffed in part by former military intelligence analysts. Based in Phoenix, Arizona, the center is tasked with gathering information on unions and social movements in order to prevent disruptions to the company's operations. According to leaked internal reports, Amazon also hired the Pinkerton Detective Agency to help the Center monitor workers, for instance, by infiltrating fulfillment centers to spot troublemakers. The firm is infamous for its role in intimidating unions and workers at the behest of late 19th- and early 20th-century industrial capitalists. Other leaks have revealed that the company monitors social media pages run by unions and other movements with a foot inside Amazon FCs, including some with a presence at MXP5.[29]

Employee monitoring is not a new phenomenon. It has been a part of industrial capitalism since at least the early 20th century. But owing to the technological firepower it possesses, Amazon is positioned at the forefront of innovation in new digital surveillance technologies.[30] The data Amazon produces by monitoring workers are black-boxed, stored in Amazon servers, and sometimes resold, making the situation even more problematic for workplace democracy. And Amazon plans to go further, investing heavily in technological development to tighten its grip on workers by expanding its digital panopticon: patents reveal the company's plans to introduce new surveillance technology, from augmented reality goggles that help supervisors identify workers, to digital wristbands that track employees' movements. Workers know that surveillance is a strategic tool used by Amazon to maintain its power. Several worker-led campaigns across warehouses in Amazon's global network have identified it as a practice to be limited or eliminated.

MANAGEMENT BY STRESS

The digitization of surveillance does not mean that worker control is fully outsourced to algorithms. Technologies and data gathered through surveillance are also used to augment managerial power. I spoke with an MXP5 team leader named Paolo who had seen worker surveillance systems from the backend. He explained how pickers walking the warehouse to retrieve commodities leave detailed digital traces for managers to observe:

> When a worker logs into the scanner they see how many pieces one does per hour [...] It's very simple, you see a line for stowing or picking, and if there is a gap in the line you can see that the worker has gone to the bathroom or taken a break.

He further described how management can see "how many pieces per hour he's doing, in which hours he was faster." Stats generated in the analysis of a picker's work can be used by team leads and managers. For instance, a manager may tell a team leader that "by noon you must make them pick 50,000 pieces." But more often, he explained, the disciplinary action was individualized. "The manager would ask me to go tell this or that worker to push a little bit because they were slow [...] I checked and then said: But they are so fast! How the fuck am I supposed to tell them to speed up? But being critical doesn't lead anywhere in there."

As with the majority of workplaces, the warehouse is a strictly hierarchical organization. This is made immediately visible through the color codes used to identify different types of workers. Those wearing a green badge (white in North America) are the most precarious, hired through staffing agencies. Blue badges are for full-time associates hired directly by Amazon. Next up the ladder are the yellow-vested "leads," those placed in charge of a small team of workers. Often, these are middle-class university graduates hired straight out of school. They typically receive some further education before being deployed to the warehouse. For instance, Amazon sends those destined to work at MXP5 to its European headquarters in Luxembourg for a training session about corporate culture and processes. Higher up the ladder still are the managers, who oversee an entire area of a given warehouse, like outbound or inbound. Amazon is explicit in its search for supervisors who "have the aptitude to train, motivate, and persuade," as stated in a job ad targeting former military personnel.

Yet the ever-increasing automation of the labor process means that many supervisors have limited technical and organizational roles, as tasks such as assigning duties to workers or analyzing inventory are mostly outsourced to algorithms—and to the engineers who write and run them. Nonetheless, these figures, which Marx called the "non-commissioned officers" of industry, command power. At Amazon, their duty is to enforce worker discipline while perpetuating the culturally constructed myth that the warehouse is a special workplace. That is to say, they are tasked with disciplining while simultaneously ensuring everyone is staying positive. Paolo summarized the dual task of managers to me: "What is the manager's role? Promising. Telling people: if you push it, I will make you a problem-solver. I will save you. Promising, baiting, soothing. And of course, punishing." Punishments can take a variety of forms: writing workers up, denying their requests for more preferable tasks, or not putting them forward for contract renewal.

The asymmetry of power in the warehouse also manifests as an asymmetry in access to information, as only managers can view the aggregated data used to calculate a worker's performance.[31] Workers are often privy to only vague figures. Managers and leads may relay things in terms of percentages, for example, telling workers that they have hit only 80% of a target, without disclosing the nature of the target. For their part, workers try to keep ahead of the disciplinary curve by attempting to estimate their productivity—pushing themselves to speed up if necessary. Some count the number of items they stow or pick per hour by estimating the quantity of items a tote can contain on average, and then counting the totes they have emptied or loaded during a shift. For instance, temp workers who hope to secure a full-time contract at the end of their stint with a staffing agency may push themselves to keep pace with the productivity rates of their full-time co-workers. In an online comment, a Californian stower described how this can work:

You're expected to meet a "rate," where you have to stow a certain number of items per week. It's not a straightforward concept, as in 1,000 items per day—each item has its own time limit/rate (something like—large items 30 per hour, medium items 60 per hour, small items 120 per hour), so it's not possible to personally calculate whether you're meeting the rate or not. Managers [...] post a list of all working employee rates throughout the day, or sometimes once per day, and that's how you can figure out how well you're doing rate-wise.

These targets vary according to the type of commodity, the area of the warehouse, and other factors depending on managerial decisions and consumption patterns.

Many MXP5 workers identify targets as a major source of stress, as workers who are not keeping pace are singled out for attention. Elisa, the young precarious employee, told me,

> Some days you can tell they have to give feedback; you see them all lurking with their little computers. That day I saw they were keeping an eye on me and tried to be fast, but [...] they called me aside and told me [...] it is my fault if we have to work overtime.

Workers from many fulfillment centers tell the same story of a team lead or manager approaching them, telling them that their rate is too low, that they need to work faster to meet the FC's standards. This relationship between associates and managers is also mediated by the barcode scanner. For example, scanners or workstation computers can be used to recall a picker or stower from the pick tower for these "feedback sessions." Elisa, for instance, was approached a number of times by a supervisor telling her things like: "You used to be a top performer and now you only reach 60–70% of the target." A target she could only guess. This indeterminacy is used to control workers. At JFK8 in New York City, termination based on TOT was rare, but workers did not know: the goal, as per leaked internal guidelines, was to "create an environment [where] associates know we are auditing for TOT," thus increasing anxiety to the point that workers would track their breaks in a notebook, just in case.[32] Workers report other complicating factors in the struggle to make rate. They think that supervisors have discretion over whom they assign tasks to. This means some workers are given "easier" batches composed either of smaller items that can be stowed quickly, or a series of items that are stored in adjacent areas of the pick tower and thus require less time walking back and forth in the aisles. I spoke with managers who denied that this takes place. And certainly, managers have an interest in maintaining that workers have only themselves to blame for their poor rates.

Discipline in sum can strike anyone who stretches out, and can come in different forms. A petty but impactful punishment that is exemplary of warehouse despotism was experienced by Elisa, who reported to me how, "To make me understand that I am no longer well-liked, they took away the only thing I have, my blue lanyard." This color code identifies instructors: rank-and-file workers who are tasked with showing new hires around the warehouse, a job which serves as its own reward insofar as it offers the occasional break from a numbing eight-hour shift of picking or packing. But workers learn how to subvert this kind of despotism. For instance, as has been commonly observed in the history of industrial work, "punitive" FC units troublemakers are assigned to can quickly become fertile ground for organizing.

At an international gathering of unions working to tackle Amazon, an American representative described this arrangement as "Taylorism from hell." And indeed, the obsession with the control and enforcement of rates creates an environment of anxiety for workers who must compete against each other and against themselves to speed up, rather than directing their discontent toward Amazon. US sociologists Ellen Reese and Jason Struna called this "management by stress."[33] The use of technology to track and quantify work, as well as to evaluate and discipline workers, is, of course, key to this form of management. A PDX9 associate explained to me how a number of things can introduce stress, like the limited availability of scanners during peaks: "You have to worry about someone stealing your handheld scanner while you are on break. It's about undercutting the workers, their ability to do the job, and forcing them to compete against each other." Emma, the MXP5 picker in her 50s, recalled a day when a lead stopped her and a co-worker to instruct them about a certain issue. Emma described to me how her colleague "kept on looking at the [manager's] computer while saying 'but I'm losing productivity, I'm losing productivity.'" Emma realized that the co-worker was looking at graphs that showed her own rates declining in real time. Though required to stop by a manager, she had not been told to log out of the system—and thus faced the prospect of later disciplinary action. The ability to watch her quantified productivity drop in real time only exacerbated the anxiety. Emma had to self-medicate to cope with stress herself: "I too had panic attacks and whatever, I would just take Xanax before work. It was necessary." Drug use is not uncommon for Amazon workers.

This sort of pressure can come from supervisors, like when Emma went to grab a snack during her shift, only to find a manager "taking pictures so they can show that you went to the vending machine outside of your lunch break." But other workers, motivated by rates or the ideological imposition of team spirit, may also participate in the surveillance of their peers. An example came from Elisa, who told me of a time when she refused a request for overtime because she was feeling unwell. "That time that we said no [to overtime] we had to log out in front of all the others who kept on working. It looked like the walk of atonement," she said, referencing a famous scene from the television series *Game of thrones*.[34] "And I was sick that day and they would go like: 'C'mon, aren't you staying?'" she continued. Due to the warehouse's need for overtime, refusal to take on extra work becomes the subject of moralism and judgment as employees internalize managerial imperatives and end up pressuring each other to keep up with the demanding rhythms of warehouse work. At times managers try to boost this collective form of control, for instance, when they summon workers to "all hands" meetings where they are lectured about the need to speed up or to put in overtime.

All of these modes of pressure are only exaggerated for workers who come to Amazon as casual laborers. During peaks, the fulfillment center doubles its workforce by including hundreds of workers hired through staffing agencies. These workers' contracts are often for only a few weeks, though some work hard in the hopes of keeping the job at the end of the specified period. Amazon managers collaborate directly with staffing agencies such as Adecco or Manpower, which have offices at the entrance of MXP5. This means casual laborers are particularly susceptible to managerial pressure, as they try to please those able to intervene in decisions about whose contract is to be extended, or who is to be part of the majority of seasonal associates that will be kicked out of the warehouse. It's unsurprising then that temp workers are described by other employees as working "absurd hours" under "psychological pressure." Some can afford it, like young male workers who are in good shape or do not have to worry about cooking or cleaning at home after their shift. It is quite different for disabled or older workers. But on the whole, a sense of stress and insecurity dominates the warehouse floor.

As in many other areas of contemporary capitalism, class dynamics intersect with race and gender and play a role in shaping power relations in the warehouse, and thus in distributing the stress in unequal ways. The division of labor in the warehouse is perhaps the most visible manifestation of how access to power is gendered and racialized, as most supervisors and virtually all managers are Italian white men, while the rank-and-file pickers and stowers include masses of women, as well as migrant and racialized workers. As put by social theorist Anna Curcio, who has studied warehouse labor in Northern Italy, "the intertwining of capitalism, colonialism and patriarchy allow companies to cut labor costs as well as to discipline and marginalize specific bodies within the warehouse."[35] Workers report managerial biases, in particular against Black and Muslim employees, and sexual harassment. Reports of favors or promotions based on sexual relationships are ubiquitous too. Elisa, for instance, had experienced this directly. She relayed that her FC instructor told her, "If you sleep with me you'll have a career." This, she said, was just the tip of the iceberg. "As a woman I have endured so many of these things." This is not unique to her experience. Many MXP5 workers report surprise at the oversexualized way young female colleagues dress and make themselves up. This is certainly not a phenomenon limited to Amazon, but following feminist sociologist Leslie Salzinger we can assume that sexualization signals a managerial culture that mobilizes women's femininities in ways that serve productivity goals.[36] More stress to be endured for the sake of fulfillment.

In North America, both the racial composition of the workforce and the dynamics of racism are different. In many FCs, Amazon's workforce is predominantly Black or brown, and the majority of working-class, racialized employees encounter racism in many forms. For instance, Black Amazon workers have been scapegoated by Amazon during the COVID-19 crisis. One of the most well-known examples is the targeting of Black American employees who were protesting against the lack of safety measures at the beginning of the pandemic in 2020.[37] In sum, the warehouse reproduces the very system that protects and magnifies white supremacy and subjects its racialized workforce to it. This was the case at BHM1 in Bessemer, Alabama, a gigantic Amazon warehouse where over 80% of the workforce is Black. As the workers strove to unionize the warehouse—a battle they lost in March 2021—

it became clear that the battle for the dignity and respect denied by Amazon's anti-union stance was not just about labor rights. It was also about racial justice, as the white upper echelon of the company fought the unionization drive led by Black workers who were first and foremost interested in being heard.

SMILE YOU'RE ON CAMERA

The coexistence of management by fun and management by stress is far from paradoxical. It is the product of Amazon's dream of broadening and tightening its control over workers. Technology is just one, albeit important, element. Another is the inherent authoritarianism of capitalist organizations. Workerist Raniero Panzieri captured this twofold mode of exploitation with his exhortation to consider "the unity of the 'technical' and 'despotic' moments in the present organization of production."[38] It is simply impossible to disentangle the role of technology on the shop floor from that of the all-too-human capitalist domination of workers. In order to maintain its economic position, Amazon does, in fact, need to find ways to force workers to seamlessly and efficiently abide by the rhythms that automation imposes. In the 1950s, philosopher Jacques Ellul spoke of a technological society dominated by techniques used in service of a dream of "absolute efficiency." In his view, technology was but one of the many techniques, organizational and technological, used to prop up such an efficiency.[39] Panzieri added a Marxist flavor when he analyzed the complementary strategies deployed by capital to subdue labor: "not only machines, but also 'methods,' organizational forms, etc."[40] aimed at organizing and structuring messy human processes, making them easier to control and dominate. Many subsequent studies of industrial capitalism have described managerial regimes like the one imposed by Amazon as the leverage used by capital to create favorable power relations on the shop floor. Burawoy called the organizational practices which regulate production in manufacturing *factory despotism*.

In the warehouse, Amazon deploys just such a set of tactics: it borrows techniques from both early industrial capitalism and the breadth of contemporary digital capitalism's cultural and technological toolkit, and bundles much of them into a single device: the barcode scanner. Digital technology provides the material basis for the management of labor on the warehouse floor, as it is used to organize work. It also augments management's ability to surveil workers and quantify their output. To borrow from Marshall McLuhan's famous take on electronic media as extensions of human social and communicative processes,[41] surveillance technology extends, or augments, management's ability to monitor employees and enforce both fun and stress in the warehouse.[42] Amazon also borrows from digital culture what media theorist Fred Turner calls a "cultural infrastructure": a set of cultural elements that can be used to structure employees' behavior.[43] In Amazon's case, this means the import of playful elements from Big Tech campuses and from the playbook of contemporary corporate culture, ranging from chocolate day, to gamification, to briefings, to wellness and mindfulness initiatives.[44]

The effect these techniques have on workers is illustrated in the web-based arcade game *The Amazon race*. It was published by the Australian news network ABC in 2019 to complement an investigative article about labor conditions at Amazon. As a player, you take the character of a picker walking around the warehouse, only earning new tasks (and thus points) for efficiency as you follow the instructions from your scanner. At times, you are confronted with the choice to stop and help a co-worker in need. But of course—that will slow you down. And failing to meet your quotas will get you fired. In the canteen, you can chat with others, but supervisors standing by the door may scold you for doing so.[45] Even in this video game, performing stretching exercises or singing with your team under the direction of a supervisor during briefings is mandatory. But at the end of the day, what really counts is your productivity. In the in-real-life warehouse, it is not that different. As put by the worker who described the use of scanners to poll employees, working in the FC is "a bit like being a peon in Warcraft."[46] This worker recognized that they are in fact a cheap and disposable unit, instrumentalized in a game being played by someone else. That someone else is, of course, Amazon's management.

The ostensibly friendly yet ultimately despotic nature of Amazon's management is not new to the history of industrial capitalism. Starting in the US during the 1920s and 1930s, corporate paternalism developed as a means of fostering the growth of industrial capitalism. Henry Ford was a key actor in this movement, as the friendly image he had built through the company's then-generous $5-a-day salary was coupled with the use of brutal repression methods. The Ford Service Department became infamous for its union-busting operations in the 1930s, quickly becoming the archenemy of the United Auto Workers labor union. Five bucks a day or not, workers described Ford plants as a "hell on Earth" that turned them into robots. Ford went so far as banning the very act of smiling (bringing about what workers called the "Fordization" of their faces), along with other forms of sociality.

A few decades later, human resource gurus like Peter Drucker updated and modernized corporate paternalism, advocating for companies to cultivate employee engagement with and participation in corporate goals. For Drucker, the enterprise "must be able to give [its employees] a vision and a sense of mission. It must be able to satisfy their desire for a meaningful contribution to their community and society." Technological companies were among the many to adopt such recommendations. In the 1980s, ethnographer Gideon Kunda studied a US company he called "Tech," whose engineers participated in a culture based on strong commitment, identification with corporate goals, and "fun."[47] Management assumed that a culture of fun could be engineered, developed, and maintained in order to facilitate the accomplishment of corporate goals. A business magazine covering the company used the headline "Working hard, having fun." Maybe Jeff Bezos read that issue of the magazine. Or maybe he simply distilled the same essential imperative for his workers from the widespread calls to build "happiness at work" found in contemporary management theory, which is prevalent in North American tech start-up culture.[48] To develop Drucker's sense of mission and belonging, the modern corporation invests in "employee engagement" techniques. Definitions of engagement vary, including workers' commitment, enthusiasm for work, and positive attitude toward the company and its values. It is, organizational psychologists insist, the opposite of the void of life that leads to burnout: a "positive, fulfilling, affective-motivational state of work-related well-being."[49] Employee engagement is, of course, designed to unleash workers' "latent capacity to do more work, to work harder," to quote critical management studies.[50] The "productivity imperative," as put by Melissa Gregg, is always at the center of contemporary managerial initiatives that borrow from self-help practices or pursue the aestheticization of work.[51] Quite different from what many workers would see as more important material rewards, such as salary and benefits, or working *less*, not more or harder.

By deploying both a new technological as well as cultural infrastructure to control the workforce, Bezos adds a digital spin to Ford and Drucker. There are differences: for instance, in the Amazonification of workers' faces, smiling is not only permitted but incentivized and rewarded as part of the culture of mandatory fun—as long as the feelings are expressed within the pre-ordered boundaries set by Amazon. Other companies follow suit. In June 2021, Canon installed "smile recognition" AI-powered cameras in its Chinese offices. The technology only lets smiling workers enter—the automation of mandatory happiness.[52] These elements, including Amazon's briefings, pizza days, and lingo, all contribute to building a paternalistic warehouse environment. In a sense this is Gramsci 101: in the 1930s, the Italian political theorist sketched a sophisticated theory of power, describing it as a process that needs both force and consent to be sustainable—a permanent and ever-changing carrot and stick. Amazon knows that despotism is not enough, and applies marketing techniques inwards, toward its workforce. After all, workers, as modern human resource theory explains, can and should be seen as *internal customers*. Management theorist Don Tapscott encouraged what he called an "extended enterprise": a company that recognizes the role of relationships in the creation of wealth.[53] The ability to extend this web of relationships both externally (toward customers or other firms) and internally (toward workers), this theory maintains, is key for a firm's ability to enhance its capacity to generate value.

Management techniques are aimed at making workers happier, but that is a secondary goal. The first intent is always to make them more productive. If they are not productive, they can remain unhappy.[54] As put by seasonal associate Zak: "The slogan is '*Work hard. Have fun. Make history*' […] I don't know about the fun part." Many quickly realize that "the firm wants to be your mom, but as soon as you slow down it turns into a wicked stepmother," to use the words of a labor unionist who had been organizing at MXP5. When they are not having fun, or even more so when they are not efficient, then the company abandons its obsession with fun and turns quickly to punishment. At a certain point, happiness ceases to be a concern at all, and the workers are simply deemed disposable and expelled from the warehouse, ready to be replaced by another wave of workers whose productivity and happiness may be more in line with corporate needs. In sum, at the core of Amazon's managerial techniques lies the struggle over the ways in which power operates in the warehouse. And at the end of the day, it is only the customers who must always be happy, their desires immediately fulfilled.

4

Customer obsession

Imagine you are ordering a birthday present for a loved one, perhaps a stuffed raccoon for your toddler nephew. It's important that you receive the toy within the next 24 hours: you are seeing him the day after tomorrow. But it's December 21! Everyone is rushing to buy toys. You are a Prime subscriber though—a paying member of Amazon's preferred delivery service. And so when you add the commodity to your cart, Amazon promises that you will find the stuffed raccoon on your doorstep tomorrow morning. You've come to expect this prompt service. Indeed, in certain cases, like when you order food items, you've come to expect the delivery might occur within a couple of hours. What you don't see is what your order sets in motion: it does not simply trigger the operations of a logistics network, but also a specific relationship with labor, resonating from customers' clicks on the website through Amazon's entire infrastructure. Indeed, Amazon's warehouses are designed around a principle of flexibility, where algorithms set a pace in response to these aggregate clicks, and workers must adapt.

The nature of online commerce means that sales fluctuate over time, for instance, with peaks in December around Hanukkah and Christmas, and on more secular consumption days like Black Friday or Prime Day. So Amazon needs an extremely flexible workforce if it is to deliver quickly, smoothly, and in increasingly unreasonable time frames, even in moments when business may double or triple. It's part of what Amazon calls *customer obsession*—one of its core corporate values, drafted by Jeff Bezos himself in 1998. Warehouse workers are constantly reminded of the slogan in briefings, as a justification for their hectic pace of work. It means that Amazon will do whatever it takes to get that raccoon to you quickly.

For customers, this obsession might mean convenience. But as the rhythm constantly changes with an algorithmic drumbeat, the workers absorb the percussive shock—the result of synchronizing the rhythms of commodity circulation with those of warehouse work. In fact, the entire logistics industry is made possible by jobs that can be quickly reconfigured, destroyed, or created depending on the needs of consumption.[1] So workers' lives must be made flexible. Amazon requires a workforce it can command to work, or not work, at will. For example, it needs workers who can't decline requests for overtime ("let's stay another two hours!") or the sudden imposition of a Sunday morning shift scheduled and communicated on Saturday evening. It requires a workforce that can double in size during certain periods, like December, and then shrink again to its baseline size in January. A workforce that can be called on to show up for an impromptu shift on a scheduled day off at a drop of a hat or a click of a button. In effect, this means that even if fulfillment centers organize work in shifts, an Amazon job can expand to affect workers' abilities to plan their days, weeks, or even the entire year. And it can do this not only by demanding overtime, but also by slowing its rhythm and constraining workers' shifts to a piddling four hours. As might be expected, the effect of this regime on Amazon's laborers is immense. As Sofia, a young MXP5 seasonal associate told me:

They take in an unmanageable volume of orders and then go, like, "guys we're in deep shit, you need to give 300% today" [...] And indeed it happens that you have to be there in one hour, only to work four hours [...] People [in the FC] last one year, not even that long. Some resist, some quit after just a few weeks, even days.

Indeed the unpredictability presses workers to adopt an always-on life that enables them to be on call at any time, waiting to work in response to any given influx of demand. Their time is monopolized both inside the warehouse and in their lives outside, as they must make themselves as flexible as possible. And flexibility is not color- nor gender-neutral. For example, since many women perform care work at home, they may find synchronizing with the warehouse more demanding, as compared to those male workers who have an abundance of time and may not mind the indeterminacy of this kind of warehouse labor.

Yet all must deal with a second type of obsession that pervades Amazon: the idea that workers can be used at will and disposed of when necessary. Jeff Bezos has called his company "the best place in the world to fail,"[2] and indeed, many fail to keep their job in its fulfillment centers. Worker failure is a recurring feature in the warehouse, as workers break down, burn out, or quit. Many are just let go. In fact, workers are hired en masse to staff the warehouse during peaks, but also fired as soon as they are no longer useful. Managers are literally evaluated for their ability to "hire to fire," as anonymous American Amazonians revealed in 2021.[3] It means that managers are incentivized to hire new workers while knowing that a certain percentage will soon be let go. In a bout of dystopian managerial lingo, Amazon calls it "Unregretted attrition rate." In other words, Amazon attaches an expiry date to the least productive workers, those who find themselves in a position where management does not regret firing them. Quite the contrary.

Workers tend to know that their employment may come with an expiry date. For some, this is explicit—inscribed in the contracts that regulate their employment, which can last mere weeks or months and bear no obligations for Amazon to tell them for how many hours or days it will put them to work. Indeed, Amazon relies on national labor laws to employ workers under precarious arrangements. These laws were not written by Amazon, but rather are the product of decades of corrosion of worker rights. But Amazon is happy to exploit them. The company employs both full-time workers it hires directly (the blue badges), and temporary workers provided by temp agencies (the green badges). In many European countries, including Italy, full-time workers enjoy permanent contracts and thus may be less impacted by flexibility requirements than the thousands of temp workers added ad hoc according to consumption cycles. While this materializes in different ways in different countries, what is common is that this precarity makes workers malleable—especially if they are contracted through staffing agencies or work in a country with low labor protections, such as the US. Their contract may not get renewed if, for example, they decline sudden demands for overtime. But even permanent workers quickly realize that their lifespan in the warehouse may not fully be in their hands. In one online comment among dozens featuring a similar tone, a Canadian employee from Brampton alerted prospective workers: "Please be warned that working at Amazon is not for everyone and that they will chew you up and spit you out because they know how expendable you are."[4] Just as other tech companies condition consumer behavior by designing their products to become obsolete, Amazon conditions its workers with the threat of obsolescence.

SYNCHRONIZING WITH THE WAREHOUSE

It may take 4,000 collective hours of work for a shift of 500 outbound associates to pick, rebin, pack, and ship the 100,000 orders assigned to the warehouse on a certain day. But this type of counting obscures a variety of more complex labor demands. Warehouse workers are not only required to provide their labor time—that is to say, the time they spend directly working in the warehouse. They are also expected to actively and continuously synchronize with the fulfillment center. Their productivity, or the number of tasks workers perform per hour (items picked, or boxes prepared for shipping), is important. But merely counting the hours they put in or the number of tasks they complete per hour does not fully reflect the nature of Amazon work. *When* they perform this work, to what extent they synchronize to the rhythms of the warehouse is as important. So, warehouse workers face a demand to get in step with the highly contingent and variant cycles of consumption that the company both fosters and depends upon.

The warehouse's hunger for such flexible work is reflected in the unpredictable scheduling many Amazon workers must deal with. Workers subcontracted through staffing agencies are the go-to resource for this, as they can be called upon or sent home with little warning. But this life-disrupting flexibility is a major concern for full-time associates too. For both, planning one's life can become difficult, if not impossible. Sitting and chatting with workers in coffee shops, they would invariably raise the problem of the unpredictable nature of work, which bleeds into their entire lives. "Normally we work the central shift, which is supposed to be 10–6, but can be 10–8, 9–5, 9–7, 9–8, up to ten hours," explained Sofia. She apologized for having postponed our meeting a number of times.

A shift change is communicated on [messaging app] WhatsApp, with a totally approximate schedule. This week we were signed up for Monday, Wednesday, and Friday, and instead we worked Monday,

Tuesday, and Wednesday. We were supposed to work tomorrow but we got a text saying that we are not working tomorrow.

Even once a shift is locked in, unpredictability looms large: "Overtime can be communicated even just 10 minutes before the end of the shift," she explained.

Overtime is an ever-crucial resource for Amazon. It grows in response to any sudden rise in orders, and shrinks if business is slow. Many workers tend to resist it, especially if it is not proposed in advance. Most labor laws and contracts only sanction overtime as a voluntary practice, and in most cases seek to limit or constrain it. Yet Amazon management presents it to workers as mandatory. Mandatory overtime, or MOT, is a major element of the labor of synchronizing with the warehouse, as Amazon uses it to sate its hunger for flexible bodies, to deal with sudden spikes or drops in orders, or simply with an increased number of trucks to be unloaded. The complement to MOT is VTO or "voluntary time off." If not all employees are needed, Amazon will send an alert so that they can choose whether they want to work that day, skipping pay or using up a vacation day if they stay home. MOT and VTO impact workers' lives dramatically. In an online comment, a Canadian associate explained it quite plainly. Amazon's schedule: "can allow for work/life balance so long as the company doesn't spring mandatory overtime." But the company does. All the time. This much is predictable: overtime will be required, over and over again, potentially, in some phases, daily. What is unpredictable is when it will be demanded, for how long, who will be asked to stay, and the impact it will have on workers' lives. As testified by a Canadian stower's review of Amazon on a website that collects employee experiences:

When the rest of Canada is having a day off, Amazon is open 24/7 and during Peak December [...] you are forced to do Mandatory overtime (MOT), it's in the contract. There is MOT all throughout the year and they can call it even if you are on a day off. So if you are enjoying a day off at a nearby town with the family, you will get a call for MOT and if you don't show up, then get attendance points deducted; there are only 6 attendance points, then fired.

Regardless of whether it is mandated or voluntary, unpredictable shift scheduling and overtime speak to the way Amazon organizes labor as part of its customer obsession. If consumer satisfaction is on the line, employees must adapt. Obviously, some workers acknowledge and even accept it. Noemi, an experienced outbound worker, told me that she understood the company's viewpoint:

> One must also put oneself in the firm's shoes. I understand that most customers do not sit down at 4 am to order something. They start from 9 or 10 in the morning onwards, thus clearly the morning shift is slow unless orders have piled up during the night […] therefore in the future there will inevitably be an evolution in shifts […] and that is determined by the market and there isn't much you can do about it.

Yet many workers are now resisting the expansion of techniques that make their labor more unpredictable, longer, and more difficult to reconcile with their lives. Workers across the company's global network of warehouses have chosen it as one of the primary targets of their struggles. At MXP5, workers succeeded in containing MOT. Italian law makes it clear that overtime cannot be mandatory, and since unions set foot in the warehouse, management may ask full-timers to put in overtime ("always with a smile," a worker told me pointing out the friendly facade of Amazon's management) but has ended the practice of simply announcing it as if it were something workers had no say in. Nevertheless, for precarious workers hired by staffing agencies, it remains difficult to refuse overtime. In other countries, all workers are still confronted with the regular imposition of MOT.

Overtime can be hard to deal with, but for some workers even regular shifts can be difficult. For instance, in early 2021, Amazon started shifting American delivery stations (its small warehouses close to final customers) to a 10.5-hour shift it called "megacycle." It begins at 1 AM and ends around lunchtime, with the goal of facilitating next-day deliveries even for orders placed the prior evening. A shift this long would be illegal in Italy and many other countries. Worker collective Amazonian United Chicagoland reported: "Amazon is setting an example for how other companies can exploit workers with this inhumane shift." Walkouts held in Chicago to protest the megacycle were so successful that managers and ambassadors had to work on behalf of striking workers. Eventually the company changed the shift's name to a less threatening "single cycle," but didn't eliminate the actual shift itself.[5] Workers are still in the fight against this brutal practice.

A MALLEABLE WORKFORCE

The individual worker's experience is radically impacted by the way in which Amazon strives to make the workforce more ductile and pliable. This was one of the first problems described to me by Luigi, the very first worker I interviewed:

> There is a core nucleus of full-time workers, but many others [...] have this contract where they work 2 or 3 days per week [...] They tell them to come on Sunday too, or not to come even if their contract guarantees three days of work: "you are not needed this week," or perhaps they send you a text message at 4 AM on a Sunday: "I told you to stay home but actually do come in this afternoon" and these are the green badgers [...] who can be blackmailed because you know, you need to come on a Sunday and show yourself, you know, contracts expire [...]

This is not unique to Amazon. Many workers employed in jobs that require them to be available on call report that they only know their shifts one week or less before the actual date. Across a number of industries, such unpredictability is commonly shouldered by workers of color and low-income workers who face barriers to full-time employment.[6] It is especially prevalent at the bottom end of the service industry, such as restaurant and retail jobs.[7] Technology has exacerbated this problem by making it easier for companies to match demand and offer—consumption and work—in an ever-faster just-in-time fashion. Think of food couriers that must be ready to deliver meals in a matter of minutes when the app they work for, like Ele.Me in China or Foodora and Uber Eats in Europe and North America, receives an order.[8]

For gig economy food couriers and e-commerce warehouse workers alike, just-in-time can not only accelerate but also dramatically slow down time by forcing workers to wait. In these instances, the labor of synchronizing with the app or the warehouse is performed by patiently not working, rather than by speeding up or adapting to the algorithmic beat. For instance, dead time spent waiting near a restaurant is a central experience for food couriers.[9] And there is lots of waiting at Amazon too. It starts even before you are employed, for instance, as you wait in lines forming early in the morning outside of a temp agency's office in the hope to be offered a warehouse gig. It continues inside the warehouse, for instance, when workers wait to begin their shift or for co-workers to finish theirs. Not working adds extra strain on workers. For years, Luigi has commuted about one hour each day in each direction to get to MXP5. But it's not just the time spent driving that edges into his non-work life.

> We carpool, three or four of us, driving one week each. The problem is they will ask your colleague to stay one extra hour [to do overtime], especially if she is green [temp worker]. So you go like "can I stay too?" And obviously they say yes, but if they don't like you they tell you to go home and you just spend one hour in the parking lot [sleeping in your car] waiting for her, or in the break room playing table tennis. Oh my god, this is your home, you can come here even when you are not working, actually you know what? Just stay always here.

And in fact, for some workers the warehouse literally does become home. In 2017, a British reporter went undercover and documented Amazon workers falling asleep in FCs: workers literally slept wherever possible, mostly standing up and leaning against things.[10] Still in 2020, Reddit users described falling asleep during their shifts due to overwork. Anna had been working at Amazon for five years at the time of our meeting, a rare holdout from among the first wave of those hired when MXP5 opened. She had seen how many temp workers, especially young men, would nap in the warehouse just to be able to put in more overtime. "The kids in there work shocking hours," she told me. "Maybe they live far away and find themselves sleeping on a couch in the break room; that's absurd." In the last few years, the media have occasionally reported that some workers were beginning to live outside in close proximity to the warehouse. In 2016, a story emerged about workers camping in a forest near an FC in Britain. In

2018, an associate who lost her home said she ended up living in a car outside DFW7, a FC near Fort Worth in Texas. In the YouTube video she posted she says: "I can't believe this is my life now [...] I work for the world's richest man and I live in my car."[11]

These may be fringe and contested cases. Yet as absurd as this may sound, Amazon has encouraged and even formalized the idea of the warehouse becoming home to its workers. For instance, the company's CamperForce program encourages American nomadic seasonal workers to live in their RVs in campsites provided by the corporation. Living across the street from a fulfillment center allows workers to synchronize with longer consumption cycles that are influenced by predictable factors such as seasonal peaks. While CamperForce is a marginal but telling example of the strategies Amazon puts in place to recruit a flexible workforce, staffing agencies perform the essential role of enrolling workers in precarious contracts. This means that seasonal workers move to the area where an FC is located to work temporary jobs.

While only employed at MXP5 for about a year, temp worker Giulia had seen this dynamic unfold and grow:

> Now we're approaching a peak and they are hiring an exorbitant number of people. There is Prime Day in July and then through January everything grows. So they come from Voghera, Pavia, Milan, Piacenza, even Bergamo, some from Southern Italy: from Molise, Sicily, they rent a room for a one-month contract.

The resulting workforce must quickly learn to staff the FC even as they can quickly be expelled once the peak ends—sometimes only a few weeks into their new job. These workers are increasingly drawn from the masses of suburban, often racialized, unemployed or under-employed workers. Without them, regardless of whether they are from Corvetto, a working-class neighborhood in the south of Milan that provides fresh workers to staffing agencies contracted by MXP5, or from Queens and Brooklyn, the New York City boroughs many workers who staff local fulfillment centers come from, it would be impossible for Amazon to expand and compress its workforce at will, which means hiring thousands of seasonal workers when needed. Many commute from far away; others must move to Piacenza or whatever area their FC is located in.

Not all workers can be totally malleable. For instance, some cannot and will not synchronize if Amazon's times clash with times predicated upon their cultural or religious backgrounds. In July 2018, Muslim Amazon workers rallied outside MSP1, an Amazon Fulfillment Center in Shakopee, Minnesota, whose workforce is composed of 30% of East African migrant workers, especially from Somalia.[12] Like other diasporized populations, Muslim communities living in the US inhabit "temporal enclaves" made of specific daily, weekly, and annual activities often dictated by religion.[13] But warehouse demands can be at odds with such religious needs. MSP1 workers faced discrimination, unable to take breaks at prayer times and unable to secure time off for religious holidays—unable to synchronize with the warehouse. Some attempted to use their regular break time for prayer, but this proved difficult: Amazon sometimes encourages workers to skip breaks in order to keep their rate up. Forced to use unpaid time off or vacation time for religious holidays, their ability to take time off for other reasons—for example, to care for a sick child—was also jeopardized. As attention to these issues grew, the protest morphed into a larger campaign organized by the Awood Center, a workers center in the Minneapolis area. The campaign forced Amazon to adjust work shifts to make them more manageable, for instance, during Ramadan, while Muslim workers fast. This early win was described as a BandAid solution by the Awood Center organizers, and came with retaliation against the most active organizers, especially Muslim women. But together with strikes over scheduling that happened around the same time in Italy and Germany, the Minneapolis struggle was one of the first instances of Amazon workers successfully taking back at least some control over time. In 2020, MSP1 workers no longer faced problems when leaving early or requesting time off for Ramadan.

WORKER OBSOLESCENCE

Amazon workers push back against the flexibility required by the warehouse. But the reality for many is that Amazon sees them as modular parts that can be easily discarded when they break down, and as just as easily replaced. The fact is, Amazon actually plans their obsolescence, as it knows that only fresh bodies can take the rhythms of work required by fulfillment and can make themselves flexible enough to synchronize with the warehouse. Apparently, Jeff Bezos believes that a stable workforce can cause a "march to mediocrity," and thus that turnover must be incentivized to minimize the presence of workers who become too comfortable or disgruntled. As reported by Amazon insiders, he thinks that all workers are inherently indolent and that their performance naturally decreases over time.[14]

As consumers, we are all too familiar with the accelerating cycles by which the technology we buy turns into waste. Planned obsolescence is a concept used to describe consumer electronics' engineered lifespan. We have all felt pressured to buy a new cell phone way too early, just because it was no longer able to deal with new apps and system upgrades. In an appropriately titled book named *Made to break*, journalist Giles Slade described planned obsolescence as a set of technological and cultural techniques used "to artificially limit the durability of a manufactured good in order to stimulate repetitive consumption."[15] In short: consumer technologies break down because they are meant to. This forces consumers to buy stuff more often than they would if commodities were made to last and not to break. Planned obsolescence is an exquisite product of 20th-century American (and Chinese) capitalism, a key component of contemporary consumerism and economic growth. As one of the biggest companies based on mass consumption, Amazon is part of this phenomenon and profits from selling gadgets with ever-shorter life spans, which need to be replaced over and over again.

There is more though, as these massive waves of replacement have been worked into more than just consumption cycles, affecting entire workforces too. In a sense, an expiry date is engineered into Amazon workers as much as obsolescence is built into technological commodities, thus multiplying precarity in the warehouse. Similar trends have been observed in other industries too. In her study of staffing agency work, sociologist Emine Fidan Elcioglu described the existence of an "organized production of precarity" that has become foundational to corporations. In their pursuit of a malleable workforce, companies may put in place multiple strategies to cultivate precarity among both their core workers, and those hired through staffing agencies.[16] Amazon takes it a step further, as precarity in the warehouse is not simply a by-product of corporate politics but rather planned in advance through a set of managerial techniques. Workers believe they are assigned expiry dates, after which they are actively let go or incentivized to quit. Work in the warehouse is indeed often temporary, even for full-timers in countries where a permanent contract means you simply cannot be fired unless something exceptional happens.

During a visit to a fulfillment center, I asked both a press officer and a manager about the turnover rates reported to me by so many workers. They denied it. "It's incredibly low, not even 1% per year," they told me. The warehouse in question had only opened a couple of years back, which may explain why many of the first blue badgers hired were still there. But this facility's numbers clashed with data from other FCs and the stories of many associates I have met. It also did not in any way account for the experiences of seasonal green badgers. In fact, turnover at Amazon fulfillment centers varies by geographical region, but on the whole tends to be extremely high. A report titled *Amazon's disposable workers*, published in 2020 by the US-based National Employment Law Project, documents the high turnover rates of Amazon warehouses in California. According to this study, Amazon relies on a "high-churn model that uses and discards workers without regard for the cost to their health or potential disruption to their lives, their families, and their communities."[17] Turnover in some Californian facilities may be as high as 200% per year, according to the report. This means that a warehouse that employs an average of 1,000 workers sees 2,000 workers being replaced every year, whether because they quit, because they are laid off, or because their contract is not renewed.

If anything, the coronavirus pandemic increased turn-over even as Amazon sought to keep up with increased demand for online shopping. In 2020, Amazon hired hundreds of thousands of new workers—175,000 in the US alone—but that did not mean that turnover slowed down. Workers kept quitting and Amazon kept terminating them. According to a *Seattle Times* analysis, in the first six months of the pandemic the company had a turnover rate at least double that of similar US employers, with tens of thousands of workers transitioning through Amazon warehouses via precarious and seasonal jobs.[18] In fact, the pandemic laid bare how much capital sees workers as disposable. Examples abound in many other companies that put essential workers at risk by failing to provide protection or by refusing to provide sick leave, and practices in Amazon warehouses across the globe demonstrated how even the risk of workers' death can be seen as acceptable if it's in the service of profit. Amazon has resisted sharing comprehensive data about the impact of the pandemic on its workforce, but it did reveal that tens of thousands of associates contracted the virus globally.[19] In Spring 2021, for instance, during the third wave of the pandemic in Canada, several fulfillment centers located in Brampton, Ontario, were shut down by public health officials to contain major outbreaks of the virus. In one case, all the 5,000 workers of YYZ4 were ordered to self-isolate for two weeks.[20] At the start of the crisis in 2020, it was only thanks to a prolonged eleven-day strike that MXP5 workers obtained personal protective equipment and social distancing measures. Unsatisfied by the changes put in place by Amazon, many elected to shirk work and avoid the warehouse as the virus ravaged Piacenza, one of the early hotspots of the pandemic. Distancing measures conflicted with the boost in sales driven by the pandemic, as many workers reported how the pressure to make rate made it difficult to observe safety protocols.

Even when not related to the coronavirus, warehouse turnover is often best understood in the context of health and safety risks. Amazon workers sustain injuries at much higher rates than in other warehouses, and push their bodies to the brink in the hope of keeping their job.[21] The painkillers distributed through vending machines are just a temporary fix. The topic of how worker bodies break down over time is frequently raised by the workers I've spoken with. Sofia had not been at MXP5 long enough to develop health problems, but recalled that: "Full-timers are those who tell you: 'I just hate them, my ligaments have gone to shit, I have to take supplements to work.'" Indeed many workers who are fed up with the warehouse's churn and burn culture or simply can no longer take the pace of the job quit in droves. MXP5's Luigi enjoyed a permanent contract that offered security against being fired, and yet he remained consistently afraid that getting out of sync would eventually make him obsolete and cost him his job. He described how "management encourages people to quit. Instead of giving you a promotion they put you in a position they consider degrading." According to Luigi, this often resulted from a perception of workers' declining health.

When they fear that you are getting old physically, that you may not be able to give what you were able to give before, and this happens around the fourth year, when you begin taking two weeks off, maybe you got the flu, that's when they start thinking "this one is suffering."

Many Amazon workers are acutely aware of the turnover rates and their causes, sometimes describing themselves with expressions like "cannon fodder." Of course, workplaces attempting to remove workers whose productivity has declined is nothing new. It was a well-documented feature of early industrial capitalism, when factory workers seen as easily replaceable could be discarded when all they had to sell, their muscular power, was no longer valuable.[22] Today, thanks to legal protections, many workers cannot be fired outright. But those workers whose performance is dwindling face managerial pressure to resign. Others would like to stay but become disillusioned and worried about their own health.

Take "long-term" Amazon employees. In Amazon terms, this means people who have been on staff for more than two or three years. An associate half-jokingly reminded me that "if you remain more than five years they give you a badge contoured in yellow, because you are a hero," as if the badge were a medal. Luigi sounded exhausted as he told me:

I don't see myself in [the FC] for ten more years, bodily and mentally. People who have been [at Amazon] for a long time look different, their eyes are turned off, they never laugh […] they have "fuck, I need to get out of here" written all over their face […] Because for the first few years you live [in the FC] full time. They renew your temporary contracts, and while this happens you can't call in sick or they'll lay you off, can't take days off. Then if after a year or so they hire you [full time] you start from scratch, and it's another year before you accumulate vacation days. On average the first time off is after two and a half or three years, and that's the first break you can take. And it's short. That's when you start being burnt out.

For many workers, the question is indeed how long their physical or mental health will allow them to continue; how long before they break or their managers push them out. As one MXP5 employee joked, "We thought we were going to retire here, but the truth is we need to quit if we want to survive until we can retire."

TAKE THE OFFER AND RUN

The truth is, Amazon doesn't want workers to reach retirement age in its warehouses. And it isn't only the slow grind described above that motivates workers to leave; in many instances, Amazon explicitly incentivizes turnover, through a series of programs that offer benefits to those willing to leave. These programs seem to allow Amazon to get rid of workers that are becoming less productive or disgruntled, so that they don't get in the way of customer obsession by slowing down or protesting the conditions of warehouse work.

For instance, Amazon provides an incentive called Pay to Quit. Known more informally as "the offer," the program encourages unhappy full-time employees to move on. In effect, it functions as a one-time cash payout in exchange for the associate's agreement to never work at any Amazon warehouse again. The payout equates to 1,000 euros or dollars per year of warehouse work, to a maximum of 5,000 euros or dollars. Years are counted at the end of the winter holiday season, which means a worker needs to endure the busiest peak of work to accumulate another 1,000. Disillusioned workers are shown the door, but not before they have sustained yet another December of hard work.

Workers tend to see the offer as a nice way to do away with them once they have passed their peak productivity, as noted by Anna:

They will try to get rid of us, I mean, what other company offers you [money] so you can quit and find another job [...] what other company tells you "if you quit we'll give you 1,000 euros per peak as a blue badge, on top of your severance package," what other company offers you money to quit?

Behavioral economists see it a bit differently than workers like Anna, noting the offer can actually function to make employees stay longer than they normally would have. Not only can it incentivize sticking around for the next peak, employees who resist the temptation to take the offer might also be made to feel more committed to their jobs.

According to social psychology, such incentives exploit humans' needs to resolve cognitive dissonance and may actually decrease turnover.[23] But that is not the reason why Anna did not take the offer. She worried about the lack of alternatives for someone like her, who had lost her professional job during the financial crisis of 2008 and was now in her late 40s. The cash was not worth risking long-term unemployment.

Other factors play a role too. For many, Amazon is a step up from other job alternatives. They see work conditions at Amazon as superior to those of other logistics companies' warehouses in the Piacenza region: MXP5 is clean and well organized, heated in the winter and conditioned in the summer. Full-time workers appreciate the steady paychecks and benefits, while seasonal workers like the ability to put in overtime and squeeze good cash out of a couple of months of work. The fact that MXP5 outsources seasonal workers to staffing agencies like Adecco is—all things considered—not worse than the outsourcing to the local cooperatives that employ most temp laborers working at the warehouses of other e-commerce companies like Zalando or TNT. They also experience extreme precarity. The rationale offered by Luigi was quite unflattering for competitors:

> I would not leave for a different warehouse, because I am better off at Amazon. People who left regret it because perhaps they now work for a cooperative, perhaps they work in a warehouse without air conditioning, without a series of things that Amazon provides [...] they regret it because they left a warehouse job that exhausts and makes you sick but is still better than others.

These tradeoffs are not always as clear cut; in other countries, workers often report that the conditions described above are inverted in their locales—with Amazon warehouses that are hot in the summer and cold in the winter, and so on.

In fact, unlike Anna and Luigi, thousands of Amazon employees globally have taken the offer and run. In 2018, Amazon reported that over 16,000 employees had used it to leave the company.[24] For the company, the point of this program is that once a worker has refused the offer, which means they passed on a chance to quit, they'll try convincing themselves they do enjoy working there. This is part of a larger trend of companies embracing behavioral economics. Amazon adopted this practice to "enable choice"[25] after purchasing and absorbing the US online shoe store Zappos in 2009. Zappos's call center employees were offered $1,000 to quit, and reportedly about 10% took the money and left the company.[26] Amazon quickly extended the program to its warehouse workers.

Another program the company has institutionalized with the goal of incentivizing workers to leave the warehouse is called Career Choice. It provides financial help to workers who want to build new skills through education. This per se is not original. Most firms institute systems of promotion that motivate employees to work hard and refine their skills, which incentivizes them to map out an entire future trajectory within the company. The difference is, Amazon does quite the opposite. The skills workers build through Career Choice are meant to help them quit the company, rather than move up the organizational ladder. On my visit to FCO1, I walked past the Career table. It was covered by flyers and signs describing the types of education opportunities that Amazon will fund. The Career program is available for Amazon associates who have worked for one continuous year and aims at upskilling them. The company claims to prepay 95% of your tuition and fees or $1,500 a semester—whichever comes first. Depending on the country, this money can be used for education in areas that are in high demand, such as "aircraft mechanics, computer-aided design, machine tool technologies, medical lab technologies, and nursing."[27] For instance, in the US Amazon uses data from the Bureau of Labor Statistics to decide what are high demand occupations. In other words, the Career Choice programs are only valid for a limited amount of certifications, licenses, and shorter vocational and STEM programs. In some areas, Amazon has also developed onsite classrooms so that some of these classes can be taught inside the warehouse. This is advertised by stressing the benefit of not having to commute elsewhere for your education. Some workers like this program. Tina, the full-time associate from FCO1 near Rome, told me that she found Career Choice to be a positive add-on to her Amazon job, if not "the only advantageous thing I am doing right now [in the FC]." In fact, the program was helping her pay for a course in accounting that she hoped to use to get a better job in the near future. Tina knew that the program's goal was geared toward facilitating turnover, adding that management "understands that after a while, employees try to get a new job [...]. It's like a car that reaches a certain amount of kilometers; after a while, they need to replace you like a second-hand car" that can be refurbished and then given away.

Normally, similar programs work for the advantage of a company because they generate an internal pool of retrained employees who can be put to work within more complex processes, for instance, by moving workers who have learned new skills to supervisory or technical positions. In some cases, a company would try to tap into skills gained by its workers on their own dime. For instance, back in 1960s Italy, the Olivetti factory workers studied by Romano Alquati dreamed of further economic and social emancipation to be achieved through education. "Many young workers study by night to try and escape their condition [and] the company helps them, albeit minimally, in the hope this would be to its advantage," observed Alquati.[28] The help Amazon provides through its program, though, is decoupled from any promise of internal mobility. Workers are to use the skills they acquire to find a new job outside of Amazon.

Practices like Career Choice and The Offer incentivize turnover, rather than imposing it through lay-offs and use of temp labor. They help Amazon get rid of full-time workers whose productivity or adherence to corporate goals are diminishing, and at the same time smooth out their exit by making it less traumatic.

In many cases, though, the fact that through both these programs and other more brutal means, Amazon incentivizes turnover by planning worker obsolescence clashes with the promise of fast moves up the hierarchical and economic ladder that workers encounter when they join the company. This generates dissonance and frustration. Indeed, Amazon does pitch upward mobility inside the warehouse: the possibility of quickly advancing into better jobs, moving up the ladder of hierarchy. This is key to the company's attempts at attracting new workers with the myth of emancipation and personal fulfillment. MXP5's Luigi recalled how the promise was made to him during hiring and reinforced on his very first day of work: "training only lasts a few hours […] and is mostly ideological. They show you how beautiful Amazon is, tell you it is an upside-down pyramid where associates— that is, us workers—are the apex, while managers are the base." So, many hope that keeping up with managerial requests will allow them to take up more desirable roles, such as lead or problem-solver—jobs that may not make a big difference for one's salary but involve less physical work and offer more varied tasks. Yet this hope is soon frustrated when workers realize that the nature of work prevents most of them from moving vertically in the organization. "They tell you 'this is a complete meritocracy, if you are worth more than the next guy, quickly you will build a career for yourself," said Luigi. "But it's exactly the opposite: If you are faster, never call in sick with a backache, after a while your back is gone, you have carpal tunnel, psoriasis caused by stress [...] and those are the first they set aside."

As this dynamic is revealed, many workers blame management for what they see as a fake meritocracy. Like other corporations, Amazon tends to hire young team leaders and supervisors straight out of a management degree rather than promoting internally. This practice generates friction as it runs against the promise of meritocracy. Often, experienced workers find themselves bossed around by new younger leads who have never had a job before and have certainly not developed any knowledge about warehouse processes and technology. The frustration deriving from this perceived injustice is only boosted by the despotic role these supervisors play. An American worker complained in an online comment, "They are allowed to audit you, dictate you, manage you. The best part, most of them are kids. They have authority and no skills relating to management at all. Their demeanor, cadence, attitude to you is the most degrading thing I've ever seen in a workplace." And yet, the most attentive rank-and-file associates know that these young supervisors are not devils. They can be even more precarious than seasonal workers, and are often left contending with the mess created by understaffing or unreasonable speed requests. They too are attracted by the myth of the relentless tech multinational, receive low salaries, and can be replaced as quickly as they are hired.

Workers themselves can contribute to this arrangement. Luigi absolved supervisors and blamed those workers who believe in the promise of a career, thus pushing their limits to achieve higher productivity rates: he called them the "bulls." He told me he advised new workers that "bulls are your enemy" and that they were doomed regardless of their speed or compliance with workplace culture. He had witnessed many waves of seasonal workers come and go: "At the beginning you quickly need to find a way to get your contract confirmed. You can be a bootlicker, or you can run. Most people run. You have to run." Complaints about a culture that favors those who pander are common among Amazon associates. Yet pandering is not enough, many would recognize. As they face the failure of the myth, workers quickly realize that Amazon sees them as disposable.

DISPOSABLE WORKERS

It might be difficult to see why any of this is good for Amazon. How does the company sustain such a high turnover rate, and why does it incentivize that? Traditionally, worker turnover is seen as a problem companies need to control or overcome. But this principle does not apply evenly across the workforce. Firms might strive to reduce turnover for those workers whose skills are seen as essential to production and specific to the company. For example, a company may present internal career opportunities or provide higher salaries to some employees. Yet when it comes to workers who are more easily replaceable, the problem may not be keeping turnover low but rather having a system in place for the company to function even in the presence of high turnover rates.[29]

Amazon has developed complex algorithmic and robotic systems to mitigate its reliance on worker knowledge. This enables the company to maintain high productivity rates, even with a young workforce. Workers become interchangeable. It only takes hours to train new associates to work as pickers, as algorithms will organize and guide their labor. In fact, Amazon training is organized around "schools," which are really crash courses for workers to learn a specific process, such as pick or receive. In turn, this permits the warehouse to rely on masses of workers who can quickly be put to work when needed and endure its work rhythms. Workers can be discarded just as quickly when they are no longer useful or productive, because replacing them is easy. In this way, both worker turnover rates and productivity can be high. This in itself is nothing new. Marx described a phase of capitalist technological development in which "the working personnel can continually be replaced without an interruption of the labor process" as "the speed with which machine work is learned by young people does away with the need to bring up a special class of worker." He observed that this allowed a "rapid and constant turnover of the individuals burdened with this drudgery."[30]

The purpose is, of course, economic. Managerial theory is brutally honest in the way in which it treats workers as investments that need to generate a return. If the projected return on investment—that is to say, the productivity of workers' skills and knowledge—is deemed to be low, firms can decide to buy labor power and skills from third parties, like temp agencies.[31] This makes it easier to quickly turnover workers. In his book on scientific management, Frederick Taylor presented the reader with "Schmidt," a fictitious German worker based on racist 20th-century American tropes which characterized German workers as dumb, submissive, and only motivated by money. Schmidt was described as being easily replaceable. For labor sociologist Harry Braverman, Taylor's writing is no less than the "explicit verbalization of the capitalist mode of production." The speed at which Schmidt learned, his receptivity to managerial instruction, and his ability to abide by productivity demands were all that management needed to account for. The turnover of workers like Schmidt was not a problem, so long as capital was able to tap into a mass of similar workers it could quickly put to work to replace those who were fired or who quit. Taylor's discussion of Schmidt is still taught in management programs across the world.

And the practice of cultivating disposable workers remains common. In her study of offshore manufacturing, labor geographer Melissa Wright described how young women cycled into and out of their jobs within a limited period of time, with factories employing a given worker "until she is worth no more than the cost of her dismissal and substitute."[32] In the factories Wright examined, this process was facilitated through a system of invasive surveillance, which involved in some cases even controlling menstruation phases in order to spot pregnancies—and fire an employee before her productivity dipped. Besides pregnant women, management fully expected that all workers would eventually decline in productivity due to injuries such as carpal tunnel syndrome, tendonitis, and back pain, as well as depression due to lack of future opportunities. Wright adopted a metaphor akin to technological obsolescence to characterize this process, describing turnover as the by-product of a process during which human beings turn into "industrial waste." Disposability was in sum the product of a calculation that measured the worth and productivity of women's labor. But as Wright also pointed out, immediate considerations about productivity were not the only reason workers were disposed of. In Wright's case studies, workers with more experience were more likely to mobilize collectively or subvert work, thus generating problems for management. Consequently, "managers face the challenge of devising a strategy for keeping [...] workers long enough to extract the value from their dexterity, attentiveness, and docility before the processes of injury, illness and anger overcome them."[33]

The application of such logics adds an ideological spin to the equation, as workers must learn and accept that they are disposable. For digital gadgets, planned obsolescence is not only about economic considerations based on capital's ability to squeeze value from the sale of more commodities. For media theorist Jonathan Sterne, "value alone is too blunt an analytical instrument here [...] A computer's social life might best be described as a kind of symbolic journey. It undergoes a series of symbolic transformations: it travels through categories from new, to useful, to obsolete, to unused, to trash."[34] Workers often undergo a similar symbolic journey. In Wright's plants, they were "produced" as disposable, that is, they were taught that they could and would be discarded, and that this was only natural.[35] In that instance, this process was heavily gendered, as management strove to instill in female workers the idea that they were naturally untrainable and impossible to upskill, and thus must consider their jobs as always temporary. At Amazon, planned obsolescence is not limited to female workers, although they may be more deeply affected by it. What it has in common with Wright's example is that it affects a profound psychological conditioning: producing workers who anticipate and accept their ultimate disposability.

The prospect of obsolescence has an effect on workers' sense of worth. As they struggle to accept this fate, they sometimes latch onto moralistic narratives to explain it. Some, like Noemi, internalize the idea that they are too weak to work at Amazon. Noemi had been working full time at MXP5 for four years, covering quite a few outbound roles. We met over an early afternoon spritz cocktail at a bar in Castel San Giovanni. She was young, energetic, and outspoken, but when asked if she would stay until retirement, she answered: "Look, I won't. Besides the physical issue, there is also a mental issue: I can be strong, I can think I am mentally strong all I want, but after a while you crumble apart [...] Everyone is quitting [...]"

In fact, many workers blame themselves for their failure to cope with warehouse rhythms. As maintained by social theorists Arjun Appadurai and Alexander Neta, failure is often naturalized as "the fault of the citizen, the investor, the user, the consumer."[36] And Amazon cultivates this narrative in its employees, telling them that their success or failure is the product of their own decisions. For instance, Amazon's WorkWell program teaches US employees that they are responsible for their well-being (and therefore productivity). A pamphlet leaked to the press in 2021 told workers that they were "industrial athletes" and provided guidelines for how they should prepare their bodies for the punishing shifts which require them to walk tens of kilometers and lift thousands of pounds. Tips on nutrition, hydration, sleep, and footwear were included.[37] Amazon plans for your exit, and still makes you feel like it is your fault. To cope with this further source of stress, many workers I spoke with described Amazon as a temporary job, a second choice, something that does not reflect their sense of self. This is especially true for white, middle-class workers with frustrated careers—a product of recurrent financial crises and waves of proletarization, as in the case of Anna. Like many others, she felt that her university degree was being wasted in the warehouse. Certainly, it had not helped her move up the ladder, although she had long given up on the idea of a career at Amazon: "This is not my job. It's only temporary [...] just the first job I found, I settled down and chose Amazon [...] but I hope to quit next year."

But some workers are trying to flip this arrangement around, emphasizing that Amazon is responsible for worker breakdown and refusing to be disposed of. Francesca, for example, was a blue badger in her 30s who had accumulated several health issues over four years of warehouse work. She told me how she joined a union for the first time in her life. Because of this, she told me how she faced pressure to quit and routine disciplinary action for not meeting her quotas. But she could not be fired legally. Asked if she will keep working in the warehouse, she replied:

That's a difficult question, you know? Actually I will, because once they broke me, relocating elsewhere is going to be hard. The job market these days revolves around logistics and supermarkets [...] and if I apply [to a job] but have to say, look I can't do this and that, clearly they'd go like well, then I don't think I need to hire you. I always say, you [Amazon] broke me, now you keep me. Unless I have a golden stroke of luck I won't leave a permanent job.

In this way, Amazon workers can sometimes weaponize their refusal to quit—pushing back on the very logic of disposability and replacement core to Amazon's business model. By deciding when and why they quit or not, they can interfere with planned obsolescence. This can become problematic for the company. In effect, worker agency in determining they cease working is a threat to Amazon's ability to manage its productive rhythms.

FLEXIBLE AND PRECARIOUS

The 2018 Black comedy movie *Sorry to bother you* imagines a dystopian corporation called WorryFree. Workers and their families are hired for life, and by signing with WorryFree, forfeit their right to quit or leave the company's premises. WorryFree workers live with their families in facilities owned by the company, dress in colorful company uniforms, and star in commercials showing how much they like the lifelong security promised by the company. Many viewers see Amazon in WorryFree. But the real Amazon hires for a limited time span. Rather than being at Amazon for life, warehouse workers know that most of them are only there temporarily, in response to both the cycles of consumption, and the cycles their own bodies go through. They are hired to be squeezed and then fired.

Still, as long as they are in the warehouse, Amazon workers must struggle to perform the labor of synchronizing with its processes. This can translate into speeding up, slowing down, or waiting—as needed. They must be flexible if they are to benefit the company's customer obsession. In fact, fulfillment centers are nodes in logistical networks that keep commodities moving. Their ability to circulate stuff ever-faster and more predictably relies on circulating people too: associates must move quickly inside the warehouse and may need to move quickly out of it as they are caught in cycles of precarity they can hardly control. As put by media theorist Sarah Sharma, they must recalibrate their lives to synchronize with what she called "the time of others"—that is, the rhythms imposed by their employer.[38] Technology is also involved, as it is used to connect consumption to warehouse processes, speed up production, and reorganize the nature of labor. For geographer Deborah Cowen, automation "works to calibrate the worker's body to the body of the logistics system" and thus synchronize the rhythms of the workers' lives to those of warehouse work and consumption patterns.[39]

Precarity was not created by Amazon. Thanks to labor laws increasingly favorable to capital, many companies in both Europe and North America have turned to intentional strategies based on mass lay-offs and temporary contracting, not only as the product of negative business cycles but even in good times. For generations of precarious workers, this unpredictability has come to be seen as something inevitable—whether it involves hopping from job to job, moving back to education, or constantly being on the job market. This is not a new condition but rather the continuation of forms of precarity that were integral to many phases of capitalism. For instance, about two-thirds of full-time US workers report working more than 40 hours per week, nonstandard schedules have been normalized, and mandatory overtime has become more and more common. Italy is not much different. And when flexibility is controlled by capital, workers end up living in a state of unpredictability. Starting with major political and economic restructuring since the 1970s, expanding with the widespread austerity measures and anti-labor politics enacted in Western countries in the last two decades, and boosted by the global coronavirus crisis, workers have been increasingly socialized to mass replacement as the normal trajectory of their working lives. This does not apply evenly across race, gender, and class lines. Men tend to work more overtime; many women must juggle with the time required by domestic work, while minority and low-wage workers employed on nonstandard schedules face a range of extra challenges, from reduced community involvement to increased health problems and even higher rates of divorce caused by precarity.[40]

If left to Amazon, this mode of organizing employment will only become more entrenched. Like the factories of early industrial capitalism, it sees workers as disposable and easily replaceable, and builds a system that allows for their quick inclusion in production cycles. In addition, it plans for worker obsolescence, attaching an expiry date to many of its employees. This practice materializes what Appadurai and Neta call "failure-by-design": In this case, the careful management of the failure of some, in order to allow others to make more money.[41] This is not a neutral process; rather, failure imposed from above reflects arrangements of power. The system comes with a heavy toll on workers, and functions only because of the laws that allow it, and the negligence of institutions that close an eye to hiring practices that push the boundaries of labor laws. Most importantly, Amazon can sustain high turnover rates because, so far, it has been able to find fresh bodies to replace those who are discarded. Without migrations, unemployment, and lack of alternatives, Amazon would quickly run out of workers.

Amazon workers struggle individually to synchronize their lives with the warehouse, but also collectively to gain more power over the organization of work; the power to influence the length of the working day, exert control over scheduling, and cultivate multi-year careers. In doing so, they challenge a system based on customer obsession and worker disposability. At MXP5, workers have managed to dramatically limit the use of mandatory overtime and are fighting to limit Amazon's reliance upon staffing agencies. In some countries, old democratic infrastructures have helped contain turnover. For instance, German Amazonians have involved local worker councils to halt dismissals and curb the practice of offering cash offers to push employees with health issues to quit. At LEJ1 in Leipzig, one of the oldest fulfillment centers in continental Europe, many workers have been employed for more than ten years, and full-timers leave at a lower than average rate. Containing precarity has positive effects on workers' lives, but there is also a political silver lining: curbing the company's ability to rely on high turnover rates means providing workers with more stable employment, which in turn allows them to accumulate the power needed to further subvert Amazon.

Technology is part of Amazon's response. Since the dawn of the industrial revolution, capitalism has kept introducing new and more efficient technologies and organizational techniques to maintain the upper hand over workers. Amazon invests heavily in developing new techniques that aim at diminishing its dependence on workers and at making ever more efficient the synchronization of their labor and lives with the rhythms of consumption. The future warehouse imagined and planned by Amazon is one where the relationship between workers and technology is smoothed out, where some workers can be replaced by machines, and where political conflicts are brought to heel by the looming prospect of increased automation.

5

Reimagine now

The Las Vegas Strip might be about as far away from Piacenza and the warehouse floor at MXP5, or from any warehouse floor really, as one could imagine. It is a place many visit to seek an escape into a heady world of neon signs, gambling, and glamor. But it's also a prominent destination for the conferences, conventions, and trade shows where businesses build hype around new consumer products and technological possibilities. For both of these reasons, it must have seemed the natural location for re:MARS, a conference devoted to showcasing the futurist breadth of Amazon's technological ambitions. Jeff Bezos called it "summer camp for geeks" when it was started in 2016 as MARS, an invitation-only get-together for roboticists, AI experts, tech executives, and futurists hosted in Palm Springs, in the Californian desert. The first and so far only edition of re:MARS, held in Las Vegas in June 2019, was the rebranded public version of the earlier, more exclusive version—open to a broader (paying) audience. MARS stands for "Machine learning, Automation, Robotics, and Space exploration," and the conference serves as a portal into a grandiose imaginary in which the warehouse plays but a minor role. At re:MARS, the public could stumble upon talks with titles like "Confessions of a CEO: Why I became a believer in robots and automation" and also gawk at prototypes of robots literally meant for colonizing Mars.

As Amazon Senior Vice President Dave Limp recounted to the audience in his 2019 talk, the concept for the event, "like a lot of good ideas, started over a glass of scotch" in a room in "Jeff's … Jeff Bezos' house." Apparently, the library in one of Jeff's houses boasts two fireplaces at opposite ends of the room. "Builders" is written on the wall above one, "Dreamers" above the other. In a sense, this image is not tremendously original: who wouldn't expect something so bombastic in the house of digital capitalism's wealthiest billionaire when we have grown accustomed to similar mottos decorating the walls of the urban coffee shops that tech and creative workers use as offices? At Amazon, continued Limp, the idea is that technology's potential is unlimited: "if we can imagine it, we can actually build it." Company executives have repeated over the years that the conference is not about the present. It is about Amazon's dreams of the future. They wrapped up their talks by saying things like, "the future is right around the corner and I couldn't be more excited," or "I am super super optimistic about the future, can't wait for 2030." Nowhere were their dreams more visible than at re:MARS, from space exploration to warehouse robotics: the technological materialization of capitalist desire.

As media theorist Nick Montfort puts it, technological future-making is "the act of imagining a particular future and consciously trying to contribute to it"—the dreamers and the builders.[1] Bezos himself made this clear during his "fireside chat." Dreamers come first, for instance, with science fiction, but dreams can't progress until builders build the technology and materialize the dream, Bezos reminded the crowd. The search for the technological fix is intrinsic to capital's development at Amazon and beyond. This leads not only to technological change, but also has cultural effects, as technological innovation is fetishized and presented as good for the whole of society—an idea that may pass as common sense in contemporary societies.[2] In Amazon's dreams, technological innovation is not simply good, but also boundless. This is something Marx also noted: capital cannot accept limits; it sees these instead as barriers to overcome. Among other limits, he famously described capital's desire to conquer, to colonize space and time. A major contemporary example is found in globalized logistics processes, through which corporations like Amazon exert control over global supply chains in real time to produce, circulate, and sell commodities. For instance, any product it sells may have been designed in urban Asia or America, made of raw materials that originate in three different continents, and manufactured in Mexico or Vietnam, to be sold in any of the countries where Amazon operates. Indeed, economic globalization based on the free flow of money, commodities, and information has allowed capitalism to expand its grasp on space and time. Judging from re:MARS, it seems that the new frontier to be colonized is the technological future itself. In one of his letters to investors, Jeff Bezos used the metaphor of a "land rush" to describe the company's approach to the internet: a nobody's land to be conquered and settled.[3]

The spectacular nature of re:MARS was part of this imagined technological conquest. As one left the Vegas heat and entered the air-conditioned convention center that hosted re:MARS 2019, a wide panoramic screen, stretching across the stage, showed alien landscapes inhabited by a lone astronaut who walked around or posed meaningfully on peaks. The swag booth was stocked with re:MARS branded giveaways, like T-shirts and plastic water bottles. It goes without saying that by tapping your badge to get the free goods, you agreed to let Amazon send you marketing emails. Beside the stage, a DJ was spinning hip hop music. The convention center, we were told, was full of participants from all walks of life: astronauts, artists, politicians, entrepreneurs, PhDs, engineers, athletes. In a 2018 MARS presentation, a speaker compared the conference to classical Greece while showing an image of Raphael Sanzio's fresco "The School of Athens"—a who's who of Greek philosophy painted on the walls of the Apostolic Palace in the Vatican, featuring characters such as Pythagoras and Hypatia. Raphael used contemporary Renaissance figures as models for the fresco, so Plato, for instance, is actually represented by Leonardo da Vinci[4]—I guess his flying machines made him the closest you could get to a tech start-up founder in 15th-century Italy. MARS is definitely not the place for modesty. This is where robot prototypes are showcased and corporate myths forged. For instance, journalistic coverage of one year's conference was dominated by anecdotes of Jeff Bezos losing a bottle flipping game to a robotic arm. Another time it was him taking his "new dog" (a Boston Dynamics four-legged robot) for a walk.

The convention attendees wore black, orange, and blue lanyards: it must be some kind of code, like the colored badges that identify associates in the warehouse. The lanyards were worn by staff and legions of men in light-colored, gingham-checked button-down shirts anxious to hand you leaflets featuring suggestions like "Your next hire should be a bot." So many men. So much light-colored button-down gingham. However, scratching the surface of these men's flamboyant speeches and visuals, it became clear that some of the spectacular innovations dreamed up by Amazon seem unlikely to materialize any time soon. For example, among Amazon's patents is a plan for a flying fulfillment center: a massive warehouse attached to a dirigible airship that can fly to a concentration of customers and station itself above them. It can float near a stadium during a football game and use drones to quickly deliver commodities—popcorn, oversized foam hands, team jerseys, whatever—to the crowd below.[5] Other patents prefigure automatic docking stations where the flying drones can be checked, recharged, and loaded with orders. In Las Vegas, Bezos half-jokingly spoke of fulfillment centers on the moon.

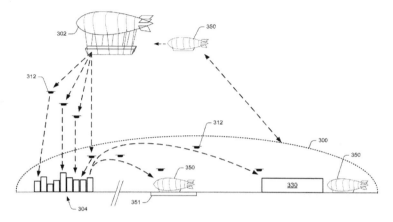

Figure 1 A flying fulfillment center uses drones to deliver products

Lunar warehouses aside, the conference was not without its doses of realness. One could wander from the fireside chat into a room where an Amazon Robotics executive explained that there will always be humans in the loop—as robots are only good for when you have to do the same thing over and over again. How could you build robots to automate the process of moving a huge sofa that is split into five pieces? he asked rhetorically. Amazon has added a lot of robots to its warehouses, but even more human workers, he noted. And so, there is a tension between the dream and the reality. Amazon is developing, prototyping, and testing all sorts of automated machinery, ranging from robotic pickers to delivery drones. Take Amazon Robotics, whose motto is *We reimagine now*. Amazon purchased this company, once called Kiva Robotics, in 2012 for over $700 million. Today, Amazon deploys over 200,000 of its robots in fulfillment centers across the globe. With it, Amazon boosted its ability to robotize and automate the labor process. On its colorful website, Amazon Robotics boasts about its "fearless resolve to achieve the improbable with real solutions." The "improbable" here means introducing "autonomous mobile robots, sophisticated control software, language perception, power management, computer vision, depth sensing, machine learning, object recognition, and semantic understanding of commands." Lots of this is indeed happening *now*, as per the company slogan.[6] But lots more is projected onto the future, as Amazon and its subsidiaries work to extend their simultaneous capacity to automate fulfillment processes and control labor.

Automation is a key component of Amazon's desire for power and control—for money, really. Automation is also seen by many as the harbinger of technological unemployment. This concept expresses a common fear that fulfillment will soon be fully automated, and the workers replaced by robots that do not strike, get sick, request overtime pay, or refuse night shifts. This specter of automation keeps haunting warehouse workers.[7] But Amazon isn't actually planning the elimination of workers from the warehouse. Despite the hype Amazon and its futurists build around automation, the company is actually much more pragmatic about the continued need for human labor. Workers will remain because they are cheaper and easier to control and discard than robots. What Amazon is dreaming of are new ways to squeeze value out of them. To treat them like robots.

PROPERTY AND THE TECHNOLOGICAL FUTURE

Understanding what corporations like Amazon have in store for us is not a trivial task. One can study things like corporate speeches and communications, prototypes, or investment decisions.[8] One can certainly travel to a Vegas convention center and try to discern how company executives imagine the future, and whether they can indeed achieve those dreams. But it is not always easy to parse out what is rhetoric, myth-making, imaginary building, and what is actually in the works— what constitutes capital's "plan," as workerists would call it. For that, there is a better object of study—one literally meant to describe the future of technology: patents. Patents are used to stake out ownership of inventions that may be developed in the future. But more than that, the existence of a patent means that a company has spent time and money to materialize an idea: it is an investment in a desired future. Patents thus offer rich material ready for analysis, including a detailed description of the invention, drawings, and references to related technologies. Furthermore, patents are public documents. Patent holders must disclose information about technology, as the invention needs to be carefully described and designed to prove publicly that it is novel, original, and useful.

Exploiting this public characteristic, I took a glimpse at the technology that Amazon may one day introduce in its fulfillment processes. Needless to say, Amazon does not shy away from asserting intellectual property rights over the technology it desires; a quick query of Google's dedicated search engine produces thousands of patents owned by the company. In 2019 alone, Amazon submitted over 2,000. That year, I spent months perusing databases hosted by institutions like the United States Patent and Trademark Office, collecting patents submitted by Amazon or its subsidiaries. Amazon owns patents covering many areas of technology, operating as it does in sectors ranging from cloud computing to digital home assistants. Out of Amazon Technologies' almost 9,000 patents awarded or submitted between 2015 and 2019, more than 1,000 dealt with inventory management in the warehouse, including robots, algorithms, and other forms of automation. In these documents, Amazon imagines a warehouse in which machines serve a managerial role, using new tools to surveil workers and boost their productivity, expanding the processes of control and dispossession already at play in today's FCs. Wearable technology and augmented reality visors capture data and provide feedback on workers' movements. Sensors analyze available space on shelves and speed up labor. Algorithms smooth out the relation between robots and workers on the warehouse floor.

A caveat: the prospect of using patents to grasp the future of technology presents a number of challenges. The main problem is that they are often deceptive or illusory in some way, and thus cannot be taken at face value. This can be incidental, but also by design. On the one hand, the technologies they describe may prove infeasible or undesirable, materializing decades later, if at all. On the other hand, patents sometimes describe technologies a company has no intention of building.

In these scenarios, the patents themselves are serviced as tools that can be weaponized in court as jamming devices, fencing off the competition with the threat of litigation should they attempt to develop a similar technology.[9] Like other forms of intellectual property, patents have a complex social life; that is, they acquire value in the different spheres where they circulate: the market, the court, and also popular culture.[10] It is impossible to discern which technologies, of the thousands presented in patents owned by Amazon, will ever be built, used, or even turned into a prototype. But they can be productive in other ways; several Amazon patents have generated media attention or wowed audiences at corporate events. Technological futures are always publicly performed, and for a reason: patents can be used to communicate with financial capital, which acts based on future expectations and predictions, and with consumers who need to be ready and hungry for the future. Finally, patents are imperfect objects of study because they present a straight problem–solution approach to the world, erasing the messiness of humanity from the picture.[11] Patents do not describe the bodies that are to interact with machines: the "human operator" or "user" is outlined in minimalist terms or drawn as a faceless silhouette. This contributes to portraying labor as falsely neutral, for example, hiding the gender and ethnicity of the workforce that staff fulfillment centers. Like other projects of Western modernity, the process of technological innovation imagines an artificially sterile future that doesn't account for, or even plans to erase, the subjects of the present, with their embodied and messy selves.[12]

Still, looking at patents owned or deposited by Amazon is a telling exercise. By their very purpose—the creation and maintenance of a private monopoly over an invention—patents show how, in the words of late British sociologist John Urry, "powerful futures" are "almost literally 'owned' by private interests, rather than shared across members of a society."[13] In its annual reports, Amazon has described itself as "an invention machine," and indeed the company invests heavily in technological development to tighten its grip on such powerful futures. As recently as 2018, it was the single biggest spender on research and development worldwide, with over $22 billion US dollars spent. By way of comparison, Google, the second-highest spender, invested $16 billion in 2018. MIT, for its part, only spends about $3 billion per year.[14]

Amazon, like other corporations, relies heavily on its ability to establish a monopoly over new technology. Economist Patricia Rikap studied thousands of patents deposited by Amazon from 1996 to 2018. The landscape she sketched is one where Amazon has moved from its early core business (a search engine-driven online marketplace) to broader areas, such as the data storage and analysis technologies that feed its highly profitable AWS (Amazon Web Services) cloud computing branch. In recent years, this strategy has further expanded to automation-related areas such as machine learning and user interfaces. The company is also secretive in the pre-patent phase, feeding off of networks of companies that collaborate with it but do not receive a stake in the ultimate intellectual property claims. Rikap described Amazon as a "predator" in the innovation ecosystem it inhabits, and as an "intellectual monopoly" in the making.[15] This dynamic can be seen in an example from the late 1990s. Amazon patented a quite broad "1-Click" system that recorded customers' credit card and shipping address information, thus reducing the number of steps needed to order items from a website. To gain leverage over the competition, the company sued the American bookseller Barnes & Nobles, which was using a similar process, forcing it to add an extra step to its online checkout process. This also meant that when Apple wanted to implement the dynamic, they had to license the patent from Amazon for an undisclosed sum.[16]

Of course, this strategy is not unique to Amazon: many corporations accumulate patents and then extract value in the form of rent from an intellectual monopoly over new ideas. As part of their desire to control and profit from technology, private interests try to expand their ownership as much as possible, colonizing the technological future. For instance, to maximize their role in extracting this value, patents can be intentionally vague and broad in scope, somewhat like the treaties stipulated by colonial powers to dispossess indigenous peoples. In effect, patent owners can use the descriptions they provide to claim ownership over large swaths of the technological equivalent of territory they have yet to even set foot upon. As in the 1-Click case, a patent for a robotic arm meant to grasp fabric may describe its components or dynamics rather vaguely, thus allowing the owner to lay claims over *any* future robotic arm designed for such a purpose.

OF HUMANS AND MACHINES

In the warehouse, in the meantime, the future is already unfolding. It was during a tour of MXP5 that I first encountered Amazon's new automated packing machine: the CartonWrap 1000. It was one of the first specimens: Amazon tested it in Italy in 2019, and then spread it to fulfillment centers across the world in the following months. Made by Italian firm CMC, the machine can pack an order every three seconds—a feat no human packer could possibly match. A huge conveyor belt, protected by transparent plexiglass, is fed with commodities. The machine then scans them, calculates the size of the box they need, sucks in and cuts cardboard, and finally wraps and labels the items. The CartonWrap's pace is overwhelming. Walking past it, one cannot avoid imagining it replacing hundreds of packers, automating out of existence a big chunk of jobs. In the news, the machine has been described as "a harbinger of automation," ultimately pointing to a future "lights out" warehouse populated only by machines. Many robots don't need lighting and a fully automated workplace could be kept dark to save money.[17] In *GigCo*, a game designed by Canadian collective SpekWork and inspired by Amazon warehouses, the player must move boxes from a conveyor belt to another. To ace the game, you need to "avoid automation," both during your shift and in the long run. Small robots that resemble Amazon's Kiva criss-cross the warehouse floor, and if you crash into one, the company introduces more to make up for workers' shortcomings. As the warehouse becomes more and more automated, it also gets darker and darker, until it is impossible for the player to see anything. That's when you lose your job: game over.[18]

Such fears about technological unemployment are not new, but rather represent a recurrent anxiety at the heart of industrial societies since the Luddites, the radical textile workers that smashed stocking frames and mechanical looms in early 19th-century England.[19] And also at the heart of science fiction, a wellspring of ideas about dystopian technological futures. Take, for example, Kurt Vonnegut's first novel, *Player piano*, set in a fully automated postwar America where former blue-collar workers are unemployed and survive in ghettos at the margins of the industrial city of Ilium.[20] The novel is often used as a parable about the future of automation.[21] The American sci-fi writer portrayed a dystopian future that, 70 years later, still seems to be one of the possible futures we face today. Unlike most sci-fi depictions of automation, *Player piano* described a world in which machines are designed by engineers but must be trained by workers. As this happens under capital's control, the effect is worker dispossession. Vonnegut depicts a machinist named Rudy Hertz as the last worker whose knowledge had been used to standardize automation processes, thus allowing capital to finally do away with human labor. Hertz is proud of having taught machines to become autonomous but also sad about his current life, which he mostly spends drinking at a pub.

Vonnegut's nightmare resembles Amazon's dream, at least partially. The future warehouse designed, imagined, and described by Amazon is one with more machines, more automation, and more systems that capture workers' knowledge and activity to improve such machines. The key difference is that this does not necessarily mean fewer workers. The future Amazon desires and plans for may be one in which labor-saving robots drive down costs, but also one in which human labor is still present. New technology will replace some workers, and at the same time make others more productive, controllable, surveilled, and flexible. Like many other corporate actors, Amazon seems to be fully aware of the continuing need for human labor, even in the face of a rapidly changing technological landscape.[22] For instance, Amazon's patents make explicit the awareness of automation's physical and financial limits, something many economists recognize. Some of the patents are so straightforward about this reality that they sound like labor sociology textbooks rather than corporate documents. A patent for a modular inventory system describes automation as:

> expensive and time-consuming to implement, unlike a human workforce, which can be allocated according to need. For that reason, conventional inventory systems continue to utilize person-nel for many [...] tasks, even though human intervention tends to increase [the] costs and decrease [the] speed throughput of any automated system.[23]

So, human workers are cheaper, more flexible, and can be replaced more easily than costly robots: they will be present even in a future robotized warehouse. But while workers and robots will co-exist, the company is concerned with making their interaction smoother, guaranteeing workers' subordination to the ever-increasingly sophisticated technology they will encounter in the warehouse of the future. At re:MARS, an Amazon executive described it as a "symphony of humans and robots working together." Interesting metaphor. Others have described a robotized metal cage that would carry a worker through the warehouse as conducting a "machinic ballet" in which human movements are dictated by automation.[24] But it might be more accurate to describe what Amazon is doing as orchestrating the increased machinic domination of its workers. As the patent makes clear, the worker's hand extends from the cage only to perform the singular part of the labor process that cannot be automated: picking the commodity from the shelf.

Amazon imagines a similar dynamic in many of its patents: machines that rely on workers to sense the physical environment and sometimes act upon it, to test solutions and teach them to algorithms and robots, and to intervene when a process cannot be automated. This entails the intensification of a form of management hungry for data from commodities, machines, and workers alike.[25] This hunger is sated through an explosion of data-generating and data-crunching technologies: many Amazon patents list a plethora of "input devices, such as pressure sensors, infrared sensors, scales, light curtains, load cells, active tag readers, etc."[26] In the future desired by Amazon, humans continue to be present in the warehouse. But their relationship with technology is changed. Machines monitor and analyze workers' activities and knowledge, turning them into data that can be used to optimize not just human labor but machinic processes too: workers train robots. Workers are also increasingly exchangeable with automated technology and intervene mostly to make up for robots' shortcomings. For instance, they act and sense for the software systems that organize warehouse labor rather than vice versa. The Rudy Hertzes of our world may still be far from getting automated out of their jobs, but they will increasingly serve the needs of machines.

CONTROLLING WORKERS, FACILITATING FULFILLMENT

The drive to increase efficiency is at the core of Amazon's plans for the future. Today's warehouse, with its algorithmic organization of labor and use of robots to speed up human work, is already pushing workers to their limits. But Amazon's patents signal that the rhythm will not slow down with the introduction of new technology. On the contrary, the company intends to introduce technology to enable the worker to bear it better. Raising productivity levels while keeping costs low is a concern for any capitalist organization, and Amazon is not different: in the future warehouse, workers must not slow down machinic operations. A patent for a visual system to assist inventory labor stresses that "performance [...] may be limited to the capabilities of a human agent performing the respective task. As the capabilities of different human agents may vary widely, processes with manually performed or assisted tasks can be subject to inconsistent performance."[27] There is also a concern that human workers may not be able to keep up with the extreme rhythms of warehouse work, which "may result in information overload or a condition in which the user is inattentive to, or ignores, the information" and becomes "fatigued or delayed," as highlighted in a patent for a new color-based interface between workers and software systems.[28] Excessive "cognitive load" that could result in "agent confusion" is to be tackled by a variety of aids, such as visual or tactile cues that reduce the amount of information workers have to deal with. These include lights pointed at the commodity to be retrieved, vibrations on bracelets worn by the worker, or arrows indicating the shortest route to a certain shelf that are layered onto a worker's visual field through augmented reality visors. Technology aims to speed up work by minimizing actions that can be time-consuming or introduce opportunities for human error, including menial tasks that are currently prevalent in Amazon warehouses, such as pushing a button or looking at a screen. In Bezos' words, the company aims to be "relentlessly returning efficiency improvements" to customers.[29]

In order to make the warehouse more efficient, Amazon plans to further automate the ways in which it controls workers. The outsourcing of managerial tasks to software systems based on data-driven decision-making is a common feature of labor mediated by digital technology, ranging from gig economy apps to online data analysis platforms.[30] The technology that is to run the warehouse, for instance, assigning tasks to workers via their scanner or controlling the movement of robots, centralizes and entrenches the power already held by algorithms in fulfillment processes.[31] Many patents describe a "central management module" or "order fulfillment system" that reminds one of Marx's automaton, the "prime mover" that he imagined controlling machines and their human organs in an automatic factory. The difference is that he may have had in mind a steam engine moving mechanical components. A patent for robotic transport units is a good example of the software Amazon plans to position as its prime mover. This "management component," or piece of software, organizes the labor of both workers and robots in the future warehouse. It:

> keeps track of the inventory holders and their locations within the associated workspace. In addition, [it] monitors inventory of the inventory facility [...] assigns tasks to the robotic drive units and other components of the system and [...] also supervises or directs manual operations, such as by indicating which items of an inventory holder are to be selected or "picked" by a worker, and where the selected items are to be placed.[32]

A number of patents detail technologies aimed at communicating to workers the result of automated decisions made by these central managerial systems. If in today's warehouse the screens of computers and barcode scanners are the main instruments that mediate between algorithms and workers, Amazon owns patents for augmented reality goggles whose function is described as "facilitating fulfillment," that is, incorporating spatial knowledge about the shelves in a software system which communicates it to pickers by generating "a visual cue or direction that is overlaid onto the field of view of the user [...] as part of turn-by-turn directions to a destination within the fulfillment center."[33]

For example, a worker wearing the visor will be shown arrows on top of her natural visual field, indicating when and where to take a turn—perhaps in pursuit of a shelf where the commodity she is to retrieve, say a coffee mug, is stored. Other wearable devices are intended to capture and analyze imaging data from shelves, generating three-dimensional models of the available space and calculating which cell in the shelf can efficiently contain an item. This would be communicated via augmented reality glasses to the worker who is to stow the commodity. The visors might even point out how to efficiently store two mugs rather than just one, perhaps by squeezing them above a pile of copy-

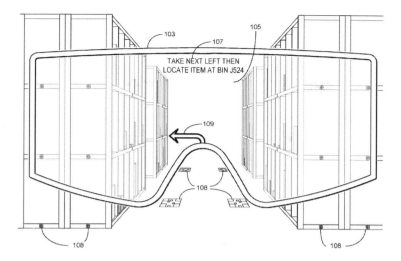

Figure 2 Augmented reality can be used to speed up labor by incorporating information about the geography of the pick tower

books. The point is "to facilitate a given task," allowing workers to store or retrieve items more quickly and increasing the overall efficiency of the warehouse. Many other patents target the intensification of human labor, but do so through relatively less "computationally intensive and expensive" technologies. One patent describes a wristband providing tactile feedback to the worker's hand. When the worker grabs a commodity from her cart, the wristband communicates the action to the central system, which in turn can make the wristband vibrate to notify the worker of a misplacement.

Centralized algorithmic control is to be coupled with more worker surveillance. The system may need to assign specific tasks to specific workers not only for efficiency purposes, but also to discipline them. Like in today's warehouse, organizational techniques and workplace despotism are inseparable in Amazon's plans, as capital must ensure that "living machines" are subordinated to what workerist theorist Raniero Panzieri called "dead machines, machines-machines."[34] Many patents make explicit the need to limit the unruliness of human workers and imagine new and more efficient ways to control them and smooth out their relation with workplace technology. For example, a patent for an algorithmic system meant to coordinate fulfillment within a network of warehouses stresses that:

> Store inventory is notoriously unreliable due to the fact that stores experience a certain amount of theft of products, or products are easily misplaced within the stores such that they are effectively lost for purposes of purchase online.[35]

To counteract this and other forms of worker defiance, patents plan for an ever-deeper penetration of surveillance techniques. The augmented reality visor described above doubles as a tool for capturing information about the worker's interaction with the warehouse. Not only can the visor determine their location within the fulfillment center, but it also "can capture imagery and/or video of items within a field of view of the user." Like many other devices described in patents, it relies on "accelerometers, altimeters, speedometers, or other sensors that can provide pitch data, yaw data, roll data, velocity, acceleration."[36] In sum, any movement can be recorded. These devices expand a logic that is already in place in the digital Taylorist processes used in today's warehouse: data can be fed to other technologies that respond to management's need to control the workforce. Among these, patents describe software that automates shift scheduling or re-routes orders in a network of warehouses if one is "unreliable," for instance, because it is inaccessible or items have been misplaced—or maybe its workers are on strike. Augmented reality is at play here too, as it can provide an "enhanced interaction system" between workers and supervisors. Imagine a supervisor wearing an augmented reality headset. When the supervisor looks at a worker, the system uses facial, clothing, or gait recognition systems to identify them. Then it projects information such as "demographic data about the user, location data within the facility, relationships with other users, messages for the user, navigation paths through the facility, access permissions" and so forth onto the visual field of the supervisor.[37] This technology has the potential to further augment workplace despotism, making warehouse workers and their labor further transparent to management.

The expansion of algorithmic control through new technology prefigures a future warehouse in which workers extend machines' ability to act in physical space. Surveillance is also augmented, as management

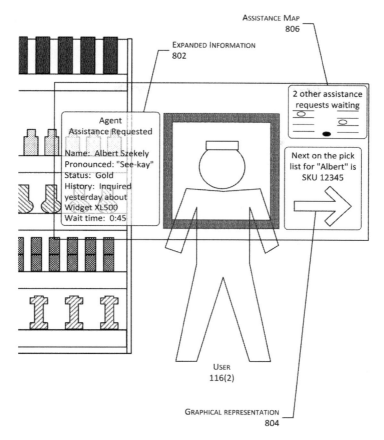

Figure 3 This augmented reality device provides supervisors with real-time information about the worker they are looking at

can access data-driven information about workers in more efficient ways. This may simply amount to the digitization and increased pervasiveness of Amazon's ability to monitor and speed up its workers. But data generated by the analysis of their activities are also used to improve machinic processes.

SENSING FOR THE MACHINE

The incorporation of worker knowledge and best practices in machinery, documentation, and organizational processes has influenced the labor process enormously for more than a century. Computing has only accelerated and deepened this process, allowing for faster and more efficient time–motion analysis and the incorporation of the results into the labor process. In early industrial capitalism, this kind of Taylorist analysis was performed by supervisors who recorded workers' movements and used this information to improve a certain process. Amazon patents work to extend time–motion analysis into the worker's body through a profusion of sensors that make data capture even more pervasive, and make it available for the software systems tasked with analyzing and feeding it back into the organization. In the 1960s, Panzieri described Taylorism as a process aimed to capture a worker's activity and objectify it "by calculating his psychic, physical, muscular, nervous abilities" and not simply their movements.[38] The future warehouse materializes this expansive description of Taylorism: in Amazon patents, data-hungry devices may capture kinematic data through accelerometers, gyroscopes, and speedometers; thermal image data via infrared sensors; visual data via optical sensors and cameras; spatial data through position sensors, compasses, position receivers and global positioning system (GPS); and more still through pressure sensors, microphones, scales, and active tag readers—allowing the pervasive datafication of all objects, activities, and interactions. All of these tools are at Amazon's disposal as they seek to record and act upon the worker.

As specified in a patent for a system that helps configure robots for transporting commodities within the warehouse, "one or more sensors may be positioned on the body of the human operator, such as integrated within a glove or other article of clothing or jewelry"[39] that can be worn or carried by a worker during the performance of any task. In today's warehouse, Amazon workers already act on behalf of algorithmic systems, for instance, when they stow or pick commodities. Amazon's plan integrates them even further with the machine: humans are imagined as carriers of sensors that extend the machine's ability to learn from its environment. They are its "conscious organs," as Marx said in his bleak prediction of a future automated factory dominated by machines. For example, through a radio frequency tracking system, the bracelet that provides feedback to workers in the form of vibration "tracks movement of one or more hands of an inventory system worker to [...] accurately identify their location in an applicable 3D space, thereby providing real-time tracking information of the inventory system worker's hand(s)." Once captured and analyzed, such information "can be used to improve efficiency of the inventory management system," that is, to improve the movements performed by workers as they store or retrieve commodities on the shelves.[40] Data capture is often triggered by worker activity, as is the case for a device that illuminates the commodity to be picked, which "may capture an image [...] when an item is stowed or when an item is picked." Sensors can capture the "head position, eye position, and/or angle of gaze of the operator."[41]

Some patents make it clear that this kind of digital, data-intensive, time–motion analysis will be used to train robots rather than to improve processes performed by human workers. Several systems aim at improving the performance of robotic arms for picking, which remains one of the most demanding and complex tasks for a robot dealing with the infinite diversity of items stored in Amazon FCs. Among the solutions imagined in patents are systems that feed on human input. In a set of patents, a robotic arm is presented with an object to grasp, for example, our coffee mug, which is not a trivial task. The robotic arm needs a successful protocol of appropriate movements, pressure, and timing to grab the mug without dropping or breaking it. Its computerized controller can use sensors to analyze the item's attributes, and search for similar items in a gripping database. If those sources are insufficient, the controller may require a worker to "generate grasping strategies," that is, to grab the mug while being subject to data capture and analytics. As described in this patent:

> Assuming that no strategies are available for this situation [...] the human operator may provide input about how the mug may be effectively grasped by the robotic arm, such as by selecting from different options presented on a screen or by donning a glove and grasping the mug so that a grasping strategy for the robotic arm may be generated using information from features on the glove (e.g., pressure sensors, tactile sensors, or fiducial markers used to track the motion of the glove with an optical imaging device).[42]

This allows the new strategy to be taught to the robot, since the "human input device" can "observe a human action for grasping an item to learn and/or determine information for forming a grasping strategy." Reflecting the task's complexity, this process is sophisticated and the grasping data:

> may include a direction from which the robotic arm is to approach the item (e.g., from above, from a side, from an angle) and/or a sequence of motions by which the robotic arm is to perform a particular grasping operation, which may include reaching the target item, grasping the target item, moving the target item to a target location, and/or releasing the target item in the target location.

Eventually, the controller evaluates the successful human output and ranks it against other strategies, thus updating the database. In this way, machinic training extends across time and space. The strategy is made available in the database for future needs, and "a strategy that was successfully implemented by one robotic arm in one location may be rapidly deployed for implementation of a robotic arm in another location in the same workspace or another inventory system having access to [the] database." Here, workers' embodied ability to grasp different objects is incorporated in software and directly used to optimize robotic processes. Interestingly, the "agent" that uses the technology described in this patent cannot be another machine. Rather, the need for a "human operator" is explicitly mentioned. Human presence in the warehouse may be progressively reduced, but robots will still rely on workers for training, maintenance, and tending.

THE PROJECT OF WORKER DISPLACEMENT

The CartonWrap 1000 does not simply pack up orders by itself. It requires human work. Like any form of automation, the machine may reduce the number of workers needed to produce a certain amount of goods. But as it eliminates packers' most dull and repetitive work, it still needs workers who load it with orders and feed it with cardboard and glue, and, like any form of automation, technicians to keep it in check. MXP5 workers call it the "bread machine," and indeed the brown cardboard boxes it churns out do resemble bread loaves coming out of an industrial oven. In Piacenza it is used mostly during production peaks. The machine is responsible for the displacement rather than the elimination of labor: new workers will perform new tasks. In fact, Amazon is still far from building its first fully automated fulfillment center. Nevertheless, the warehouse is continually built (and imagined) as a workplace in which the relationship between workers and machines keeps shifting in favor of the latter. This won't happen overnight, but it is an incremental project. Technology is another tool used to make workers obsolete, but most patents that aim at radically changing the technical organization of Amazon work are currently based on the coexistence of humans and robots on the shop floor.

The main exceptions to this rule may be in last-mile delivery, an area that occupies a major position in Amazon's innovation efforts. The company owns hundreds of patents for drones and other automated vehicles for home delivery. In the last few years, it has been building and testing prototypes of such robots. Its Scout, a blue six-wheeled electric robot carrier, is already zipping on the sidewalks of a handful of American cities. More consumers may in the near future interact with robots tasked with delivering from the nearest distribution center to their doorsteps. These robots need human intervention, be it for maintenance or because they must be operated remotely, at least when something goes wrong. Yet even if they don't destroy jobs, they may contribute to the geographical displacement of work. The crowdwork platform Amazon Mechanical Turk already allows the company to displace labor overseas. But unlike data tagging or customer service, fulfillment centers must be positioned near affluent urban markets and cannot always be relocated internationally to find cheap workers (nor are they to be found on the moon). But what if the human labor required to operate warehouses in Piacenza or New York was performed from Colombia or the Philippines, seizing the advantages of more deregulated labor markets? Amazon owns patents for robots that can be operated remotely. This is not new: just think of the surgeons performing operations at a distance through a remotely operated robotic system. Or the drone pilots who bomb the Middle East from an American office—some are actually based a few miles outside of Las Vegas. Now imagine the same principle applied to a warehouse: a robotic picker in MXP5 that can be operated by a worker sitting in a different area of the world and connected to the warehouse through Amazon's digital infrastructure. In a patent for such technology, the worker uses a virtual reality headset capable of receiving images from the robot and a joystick or a pair of sensor gloves that turns the movements of their hand into input for the robot. As the worker sees a virtual shelf through the VR headset, they grasp the commodity the robot is to pick, say a stuffed bear, and the robot's mechanical arm reproduces the movements in the warehouse. The gloves can, of course, provide tactile feedback to the worker, making their labor more realistic.[43] This technology would materialize something imagined in *Sleep dealer*, a 2008 sci-fi movie directed by Alex Rivera. In the movie, Mexican workers are hired by a sweatshop in Tijuana to remotely operate robots working in construction in the US. The result is that America can import cheap migrant labor from another country without the hassle of importing migrant workers.[44]

Back inside the warehouse, for now Amazon seems to desire a workplace in which humans and robots are interchangeable. In fact, many patents do not state who or what interacts with the technology they describe, and use words such as "entity," "agent," "user," or "operator," which "may refer to a human person working in the materials handling facility or to an automated piece of equipment configured to perform the operations."[45] In this way, Amazon is setting itself up for a future in which either can be deployed. To achieve these goals, Amazon patents imagine ways in which machines can work to recursively improve themselves. These machines not only benefit from but also are the subjects of Taylorist feedback loops of time–motion analysis and corrections. For example, in a patent for robotic technology that improves the warehouse's ability to store commodities in limited space, moving shelves around to squeeze more stuff in a limited area, the software system can analyze robots' movements, rank them according to their efficiency in storing the shelves, and then feed the results back to robots, instructing them to perform the movements in the most efficient way.[46] This process can be decentralized and assigned to single robots if it is putting too much stress on the central software. Like humans, the warehouse's "automaton" in charge of controlling several components of fulfillment processes can suffer from cognitive overload, and so the robots themselves may be asked to step in and take up "decision-making relating to certain aspects of their operation, thereby reducing the processing load on [the] management module."

Even within a future imagined automated workplace, several patents take for granted the continuing need of labor's presence and plan for a warehouse floor in which the cooperation and coexistence of humans and robots must be facilitated. At a basic level, this means imagining an environment in which robots can, for instance, sense RFID chips worn by humans who invade their space, perhaps to conduct repairs, and thus route around them to avoid collisions—Amazon is already experimenting with vests that keep robots at bay. Other patents acknowledge more explicitly the physical and economic limits of automation, for example, describing how a certain task can be assigned either to a human or an automated operator, or simply to "other components" or "[an]other suitable party" of the inventory system. Indeed, fulfillment processes described in these patents tend to be designed as flexible or "modular." For instance, sorting stations can be staffed by either humans or robots, and different kinds of sorting stations or conveyor belts can assemble and reassemble in different shapes to accommodate the flexible deployment of workers or robots. Another patent that accepts the inevitability of human labor in the warehouse aims to "facilitate the division of inventory item processing," which can easily be translated in division of labor, "between automated and manual options [...] human operator or robotic manipulator."[47] In this case, the patent describes a system that determines whether a certain commodity that is to be picked or stowed can be manipulated by a robotic arm. If not, the robotic shelf will be dispatched to a section of the pick tower staffed by humans. Human operators will thus be called in to complete tasks that are not suitable for robots, for instance, when machines are unable to grasp a certain object due to its shape, weight or fragility.

As workers are performing tasks assigned by the machine, they may need its support. Anything to smooth out the relation between workers and robots. To do so, Amazon works to automate the labor involved in worker support and assistance. Can the machine interpret and interact with humans' emotional states? Several companies are working toward forms of artificial intelligence that aim at recognizing emotions in order to perform exactly the kind of relational labor involved in listening to and taking care of humans.[48] For example, Facebook owns patents for a "boredom detector." By analyzing a user's clicking or tapping patterns, as well as observing their facial expres-

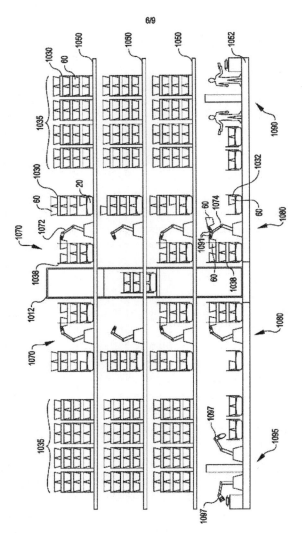

Figure 4   This system automates the division of labor between workers and robots

sions, Facebook's algorithm tries to identify whether they are getting bored and about to stop interacting with the platform. If data points out that the user may soon disconnect, the system changes the content shown to them. This is based on forecasting about content's ability to generate interest in the user, and thus keep them online.[49] Amazon's version of this kind of automation is described in a patent entitled "Using gestures and expressions to assist users," relevant to both fulfillment processes and the automation of grocery store services in Amazon Go or Whole Foods stores. Rather than boredom, this system aims to detect frustration. Imagine a worker walking in the pick tower to retrieve a certain commodity that has been assigned to her through their barcode scanner. A set of imaging and spatial sensors capture her position in space, movements or facial expressions. If she can't find the coffee mug:

> The user may be walking up and down the aisles scanning [...] presenting expressions illustrating that they are having difficulty locating an item. For example, one or more microexpressions illustrating frustration may be presented by the user and detected by the inventory management system.[50]

If frustration is detected, or perhaps we could say algorithmically calculated, Amazon's system will send an assistant to ask, "How can I help you?" Once again, the patent leaves the nature of such an assistant open. Someone will be dispatched to the location of the user to provide assistance, but "the associate may be a human or robotic system." By automating this kind of work, this system plans for a technological future in which machines replace forms of service labor that are traditionally assigned to women rather than more masculine warehousing operations.[51]

Whether or not the patents can be read to suggest that a fully automated fulfillment process is the ultimate goal of Amazon, most work toward a near future in which a flexible workplace can shift tasks between humans and robots. The implications run against the idea of a future widespread unemployment caused by automation. Indeed, for the time being, things seem to be going in the other direction: even as automation is increasingly deployed in existing fulfillment centers, Amazon keeps on relying on masses of workers performing physical jobs under algorithmic control. The same is true for other companies that work to automate more and more warehousing processes. For example, the UK online grocery store Ocado operates warehouses that store commodities in "hives" composed of cells served by robots not dissimilar from Amazon's Kiva. They retrieve items and group them in grocery bags ready for shipment. Videos of the warehouse show a chessboard of moving robots and a maze of conveyor belts moving totes around.[52] Human pickers are aided by robotic arms: some are suction-based and can pick tins or boxes. Others mimic a hand and are used to pick irregularly shaped items. All need human intervention, and Ocado workers, too, are subject to widespread surveillance and to the strict enforcement of work rhythms dictated by the robots. More automation has not necessarily made their work easier. At Amazon too, new waves of robotization will change patterns and types of work. Human labor will not disappear but will be used to improve robotic processes. This generates interdependency between machines and the organization of work, and the introduction of robots makes human interaction with those systems more crucial, rather than removing humans from the picture.

SPECULATIVE SHIPPING

The future of workers may still be unclear, but certainly Amazon has plans for the future of consumption, too. Consumption is an all-too-human endeavor, capricious and ever-changing. It must be tamed. Like other companies, Amazon strives to forecast changes in consumption patterns, such as short-time peaks, which require more workers to be active in the warehouse at a given moment. It really is about predicting (or trying to predict) the future, that is to establish which exact commodity or type of commodities may be in hot demand at a certain time and place. So, Amazon relies on forecasting algorithms and other methods to determine how likely it is that a certain item will be ordered, as well as when and where. Some techniques are rather unsurprising. For instance, weather forecasts inform predictions about spikes in orders: if next Sunday will be rainy, more people will spend the day at home in front of their computers or phones, and thus order more stuff. If Amazon predicts that many orders will pour in, it can plan to have the workforce necessary to process and ship them the next day. In this case, more workers will be needed to staff Monday shifts. Other, more sophisticated calculations are based on the extraction of data, for instance, from user activity on Amazon's websites. These data are algorithmically analyzed and then used to organize the circulation of both labor and commodities in and out of the warehouse. As no fulfillment center can contain all the commodities customers can purchase through the website, Amazon may decide to move stuff around its interconnected network of regional warehouses and local distribution centers if it thinks they will be purchased in a certain area. This is something that Nico, an Amazon manager I talked to, described to me:

> So you click [on a commodity] and they tell you that you can receive it tomorrow or the day after. Perhaps this item is shown [on Amazon. it] but it is not in the warehouse yet, perhaps it is in Germany [...] Based on how many people are viewing it in Italy, through an algorithm, when they see that one hundred or one thousand people in Italy are viewing Converse shoes they start shipping those shoes to Italy. Thus there are trucks that keep circulating.

The immediate consequence faced by workers (and managers) if forecasting is correct, Nico continued, is that a truck arriving in Piacenza "may contain orders that have already been placed, and therefore [it must immediately be made] available for the picker" by quickly unloading, scanning, and stowing the commodities it contains. In fact, "speed is everything" because "you can't store every item in a warehouse, but when you promise a delivery and in fact actually fulfill it, what moves in the backstage is over the top," Nico explained.

Amazon is continually trying to forecast demand even at a more granular level. For instance, the company owns a set of patents for "anticipatory shipping" or "speculative shipping" algorithms: here the race for speed literally bleeds into the future.[53] This would be a method to deliver orders even before customers have clicked "buy." The patent states that one substantial disadvantage of e-commerce is that "customers cannot receive their merchandise immediately upon purchase, but must instead wait for the product to be shipped to them." Therefore the algorithm tries to anticipate or calculate the chances that today someone in a certain neighborhood in Milan will order a certain commodity, say a pack of toilet paper rolls. The item is boxed and shipped to the geographical area but not addressed to a specific customer yet. Only when (and if) someone orders it online, the system attributes an address to the package and then delivers it to the customer's doorsteps. The technique would thus not only decrease the chances that a customer would switch to a brick-and-mortar store or to a different e-commerce company, but also help Amazon manage fluctuations in its need for warehouse workers.

The use of the definition "speculative shipping" in this patent is significant. Many digital platforms rely upon predictive algorithms that use data analytics to forecast future outcomes. Social media aims to calculate probable future outcomes via the analysis of social interactions, and the platforms then use these predictions to make decisions in the present. For example, Instagram needs to analyze my interactions on the platform to produce probabilistic inferences as to whether, around lunchtime in Toronto, I will click on an ad for an Italian restaurant rather than on one for a Canadian greasy spoon diner. Speculating on future outcomes based on probabilistic calculations of risk: these algorithms sound a lot like the financial market, and in fact finance uses big data analytics to infer (or imagine, as social theorist Louise Amoore puts it) "a range of potential futures" that can be used to calculate the risk associated with an investment.[54] In Amazon's case, the risk is represented by the resources and money spent to ship a truckload of shoes across the border or drive a box around town without knowing for sure if and when they will be purchased. Algorithms calculate this risk and provide the company with the information it uses to speculate—that is, to decide which commodities should be moved around and when.

WHOSE FUTURE?

Through its work-in-progress of reaching increasing levels of automation, Amazon develops new technical foundations that consolidate its power in the digital workplace. Task standardization, algorithmic management, surveillance, and time–motion analysis prefigure a future warehouse in which human labor operates as a new kind of appendage of machinery, making up for its shortages. Through myriads of sensors and other devices, workers act and sense on behalf of technology, extending its reach inside the warehouse. If Amazon sells or licenses its patents, its future technologies may be used by other corporations, although many are already working toward similar innovations.

The notion that humans will one day extend machines can be traced back through the history of computing, Jesse LeCavalier described in his book on Walmart's warehouses.[55] LeCavalier went back to a foundational 1960 essay on human–computer interaction written by J.C.R. Licklider, the American psychologist and computing pioneer who first imagined forms of artificial intelligence that could analyze people's "mental work" and thus extend and complement humans' cognitive capability. This was to be a symbiotic relationship between human and machine, pretty much in line with Marshall McLuhan's famous idea of media as extensions of human senses and organs. In this logic, a television set extends our ability to see images at a distance and asynchronously. But Licklider also posited an opposite, dark side of future human–computer interaction: a form of automation that "started out to be fully automatic but fell short of the goal,"[56] one in which human operators serve the technology rather than the other way around. This he called "humanly extended machines." The technologies deployed by both Walmart and Amazon align with this latter approach, as they imagine human workers in the warehouse as extensions of machinery's ability to perceive, learn from, and act upon the environment. Workers thus come to constitute "an organic extension of the computer systems that control the environment but lack the dexterity and cost-effectiveness to execute the commands," as put by LeCavalier.[57]

The idea of humans extending automation may sound dystopian, but the utopian hype spread in places like re:MARS must also not be taken at face value.[58] In the early 1960s, workerists cautioned against the persistence of "myths" about the role of technology as a natural and progressive force. As pointed out by feminist technology scholar Judy Wajcman, even the very language used to describe automation ("neural networks," "intelligence," "learning") is dense with misleading anthropomorphic metaphors that aim to make technology appear as something natural, thus obscuring its contested social and political nature.[59] These myths need to be dismantled if one is to understand the way in which capital uses machines to subdue workers.

Of course, we do not know whether the algorithmic and robotic technologies desired by Amazon will ever materialize and enter the warehouse floor, or in which shape, although we cannot rule out the possibility that some may already be in place. What we can safely assume is, as one patent puts it, that "it is inevitable [...] that the paths of the robots and humans working in the warehouse will cross." The authors of this patent are concerned because "direct contact between the human workers and the robots [...] can be problematic, and a maintenance issue for the robots."[60] But problems may arise for workers too: what about the maintenance issues they will encounter as they face ever-innovating forms of inequity and are increasingly subordinated to machines? From this viewpoint, technological evolution serves the purpose of entrenching capital's power over workers rather than being the result of an advancement of knowledge. There is no objective factor implicit in the development of technology under capitalism that guarantees a radical transformation of social relations. In sum, automation is not meant to liberate humans from work under capitalist relations. Rather, it can perpetuate and consolidate the authoritarian organization of work. Workers are concerned: "At this point they may as well hire actual robots," reads a comment left online by an Amazon warehouse associate from Florida.

But one does not hire robots, at least not in the near future. The patents owned by the company seem to anticipate a future warehouse in which some forms of labor are made obsolete, but masses of human workers are still needed. We can imagine them attending to robots, performing tasks decided and controlled by distant algorithmic systems, wearing sophisticated sensors that feed information to machines, and being subject to even more pervasive surveillance systems. They would work under machines, not with them: a historical reshaping of the conditions of production under capital's new ability to capture workers' knowledge and organize their labor at a planetary scale.[61] At the same time, their number could be reduced as automation increases the efficiency of warehousing processes. This will prove useful to Amazon should the corporation's current ability to access a reserve army of cheap human labor diminish.

Does it have to be this way? If people are still working, they will bring to the warehouse their imagination, their new desires. Urry reminded us that our technological future is "neither fully determined, nor empty and open."[62] This has to do with the political nature of futures, which are contested and saturated with material interests. A recent wave of

feminist and Black technology studies on robotics and automation proposes that the future of technology is "predetermined by techniques of differential exploitation and dispossession within capitalism" and thus deeply intertwined with class, gender, and racial structures.[63] Of course, new technology can contribute to alleviating such structures of inequity as well. To give one of many possible examples, even in the face of its deeply exploitative nature, the gig economy's low barriers to entry have been cited as increasing access to the labor market for diasporic and racialized communities, as shown by Uber's popularity in French *banlieues*.[64] More often though, technology reproduces or even reinforces structures of inequity, as in the case of the persistence of sexist and racist stereotypes in Google search results.[65] What these examples have in common is that, when left to capital, technological change is always geared toward domination and profit. Neither executives' speeches at re:MARS nor the patents written by Amazon's lawyers and engineers seems to be concerned with much else than that.

The picture is bleak. In fact, a paradox of digital forms of automation is that the indispensability of humans does not prevent such systems from actually working against the majority of human beings. But technology is a human artifact, and many forces can drive its evolution or influence the results of its deployment before the moment when the portal of the future is shut. Certainly, the nature of the task being automated, regulation, or the cost of labor matter. Power relations between labor and capital are even more important. The future is not just a top-down imposition: it can be struggled over. Capital's plan, in sum, is not omnipotent. So if we do not like what the builders are doing, can we at least replace the dreamers? If the future is never fully written, then the power balances of tomorrow's technology are in workers' hands too. In Vonnegut's *Player piano*, humans plan to revolt against the fully automated capitalist dystopia they are forced to live in: "Those who live by electronics, die by electronics. *Sic semper tyrannis*." For some years now, Amazon workers have contested and resisted the ways in which corporate technology shapes their lives. They dream of a better world, and at the same time organize here and now to achieve it. Their struggles may help us reimagine a different future, a new path of liberation from the grip of digital capitalism.

6

Make history

It's early 2021 and MXP5's parking lot has never been so busy. Not because of a seasonal peak, but because it has become a stage for worker demonstrations. On March 8, it was feminist collectives and workers from other warehouses celebrating International Women's Day and demanding better working conditions for women employed at the fulfillment center. Just two weeks later, it was the local picket line for a national strike that involved Amazon's entire distribution chain, from fulfillment centers to couriers.

In the wake of the 2017 strike at MXP5 and subsequent strikes in other European FCs, Amazon has become a global symbol of a renewed clash between labor and capital, as workers have set themselves up as a check on Amazon's power. Yet this fight is not merely symbolic. Far from it. Under banners such as *We are not robots* or *Strike hard, have fun,* workers across the world are protesting their work conditions, targeting the systems of worker surveillance that underpin the warehouse, work rhythms, and the lack of safety on the job, especially during the coronavirus crisis. They have asked for better pay and benefits, more power over scheduling, and a stop to precarious employment. Soon, they might also start to address the scope and purpose of automation on the warehouse floor.

In many ways, Amazon has become a symbol of a broader threat to the labor movement for one simple reason: other companies, across a range of industries, are increasingly emulating the techniques Amazon has pioneered. The successful Amazonification of other sectors of the economy in this way would be a defeat of historical proportions for the labor movement, as it would legitimize and spread the strategies used by Amazon to control the workforce and squeeze labor out of it before discarding it. Indeed Amazon does not simply neglect workers' livelihoods and dignity; that is a common feature of capitalism. Rather, it aims at something transformative. In reinventing century-old logics

derived from early industrial capitalism by augmenting them with digital technology and new managerial techniques, the company is building a new form of exploitation in the service of its economic goals. Thus the fight against such a powerful opponent has the potential to be era-defining, transcending questions about the conditions of work in any individual warehouse. As was true for the industrial workers of the 20th century, worker-led struggles in the most advanced sectors of today's digital capitalism are likely to have ripple effects that define work relations in other industries. As one of the highest "development points" of capitalism, Amazon is the place where "the subversive strength of the working class" has strategic potential, to cite Panzieri.[1]

So it is only fitting that workers have appropriated one of Amazon's core slogans into a proclamation I've now heard at many union meetings: "We are gonna make history, we're gonna fight, and we're gonna win." And a history made by workers would be very different from the one Amazon is writing. A history from below can reveal the social relations hidden behind the smiling arrow. Without their mobilizations and testimonies, the reality of the warehouse would remain opaque. When I started researching this book in 2017, Amazon was hardly on the political map. Today, one only needs to open a newspaper to read about the company's labor politics. It is now widely acknowledged that the Amazon warehouse is a laboratory where a new relation between capital and labor is being tested.

Amazon uses technology to dispossess workers from precious knowledge that previous generations of warehouse workers have leveraged tactically in their fight against exploitation. It hones managerial techniques to boost productivity, incorporating psychological strategies like gamification, and enforcing a specific workplace culture of fun. It equips its supervisors with a pervasive system of surveillance that monitors workers and can be used to isolate, threaten, and punish them. It exploits, if not abuses, labor laws to make the workforce precarious and easily replaceable. It follows a corporate anti-union playbook to prevent organizing in its warehouses. It engenders high worker turnover, pushing out workers it deems obsolete if they can't or won't abide by its requests and rhythms. Finally, it develops new technologies that will enable it to deploy each of these tactics in ways that are more efficient, fast, and pervasive in the warehouse of the future. These are the components of a conscious effort at decomposing the

workforce: individualizing workers and preventing them from coming together and recognizing their collective interests are diametrically opposed to Amazon's.[2]

And this offensive against labor does not stop there. As it realizes that climate change and environmental degradation signal that capitalism is running up against limits imposed by planet Earth itself, Jeff Bezos dreams of his own planet B. Blue Origin, a company he founded, is developing spacecraft it hopes to land on the moon in 2024. The goal is to create a permanently inhabited station on our satellite—perhaps a fulfillment center too—and then use it as the springboard to one day reach and colonize Mars. Blue Origin's slogan is *Gradatim ferociter*, which is Latin for "gradually, ferociously," echoing Amazon's own self-imposed imperative: "relentless." Bezos' ferocious search for the new frontier in space continues the dream of a *terra nullius* that can be colonized, where capitalism can expand forever, frictionless. The same fantasy is being pursued by Tesla's Elon Musk. Other Silicon Valley billionaires, like PayPal founder Peter Thiel, have imagined building artificial island nations in the Pacific Ocean, free from government interference.

Down on Earth, though, things are more complicated.

In defiance of Amazon's efforts to fragment them, workers are coming together: a number of unions, worker collectives, and other worker-led organizations around the world have established footholds inside Amazon's warehouses. They are joining a fight for distributive justice, arguing that the nearly infinite amount of money accumulated by the corporation should be redistributed. But it is not just about the money. The fight against Amazon is also a fight for racial and environmental justice, for health and safety, for workplace democracy, and for worker control of data generated by labor.

The current conditions of work in Amazon warehouses are the product of more than the decisions of a single company. Numerous factors have enabled this. Without the reforms that have liberalized the labor market, financial globalization and the legacy of colonialism, the influence of corporate power over local and global politics, the laughable corporate environmental responsibility practices,[3] the intellectual property laws that privatize innovation, and the unjust systems of oppression that intersect dimensions of class, race, and gender, Amazon would be very different—or perhaps it wouldn't

even exist. That said, the company's choices have tremendous consequences, including its decisions on technology. When led by capital, technological evolution is driven by the desire to increase productivity and subdue the workforce. The warehouse strives to make workers more machinic, turn their bodies into easily replaceable spare parts or interfaces for the machine—if not to turn them directly into robots. But workers resist and subvert complete subjugation by the machine. A different relation between humans and machines, between workers and robots on the warehouse floor can only be the result of workers' active participation in shaping the machine itself. Amazon dreams of a frictionless flow of commodities and money. Workers' resistance may reshape or even stop this plan.

ORGANIZING THE WAREHOUSE

The fight against Amazon is uphill. The culture of the warehouse is infused with anti-union rhetoric and an imposed mythology of individual success and failure, its workflows are shaped by techniques that atomize workers and preclude interaction, and workers are constantly surveilled for any signs of political activity, and disciplined when it is detected. That does not make it easy for workers willing to take up the fight. As recalled by Lisa, one of the first drivers of organizing efforts in Piacenza back in 2017, unionizing MXP5 was not easy. For most of the handful of associates who invited unions into the warehouse, it was their first experience of labor action: "At the beginning it was pure terror," she told me. "The first few [card-carrying members] were considered walking dead. To journalists asking why there was no union [already] in such a big firm, [management] gave an answer straight from the 1950s: there is no need for a union, because it is us who best protect workers." This rhetoric is at play beyond Italy. When faced with the threat of unionization, management in FCs in other countries has stressed that having a union would hurt innovation, and thus worsen prospects for workers. Better to rely on the value of a "direct working relationship" between the mega-multinational and its employees, as Amazon has put it in training videos aimed at its supervisors.

In fact, the company is constantly working to be union-free in all the countries where it operates, and at all levels of the workforce. Seattle, for example, is one of the last strongholds of the American labor movement, and yet Amazon, one of the biggest employers in town with the tens of thousands of workers employed in its offices, is completely union-free. Amazon is willing to go to great lengths to prevent unions from setting foot in its warehouses too. While surveillance practices and the reliance on easily discardable seasonal labor are now well established, more mundane tactics are not to be overlooked. In early 2021, as workers from BHM1 in Bessemer, Alabama, were campaigning for the first-ever unionization vote in a North American FC, management created its own anti-union propaganda campaign, including setting up a website, plastering even toilet stalls with anti-union flyers, and mobilizing worker "ambassadors" to express anti-union sentiment on social media.[4] Union organizers even claimed that Amazon managed to get the city of Bessemer to shorten the time of a traffic light near the warehouse, making it more difficult for workers to hand out flyers to colleagues arriving at work. A seemingly desperate move maybe, but it speaks to the power the company has to influence local politicians. In Piacenza, this same kind of last resort technique was even more explicit: in 2017, management denied MXP5 workers a room for their first union meeting, forcing them to hold it in the bathrooms—which did not stop them from organizing their first strike a few months later.

Episodes like these can lead to more determination in workers, and in fact end up becoming part of a David and Goliath mythology of fighting back against an all-powerful opponent. In the last years labor unions have set up shop inside Amazon warehouses, including in Spain, Germany, the United Kingdom, France, and a number of other countries.[5] But in Europe, Amazon workers organized from the bottom up have immediate access to organizations ready to support them. Other countries are further behind. In Canada and the US, where unionization must be won through a vote, there has not yet been a successful effort. Many around the world watched as in April 2021 the unionization drive at BHM1 in Alabama came close, but was ultimately defeated—at least for now. Just a few weeks later, in May, MXP5 unions won a historic vote as they elected the first formal representation in an Amazon warehouse in Italy, reaching the needed 50% quorum. Regardless of formal representation in FCs, unions and worker collectives worldwide participate in global networks that coordinate actions against and research about the corporation. All have formed to take the fight at the transnational level—the only possible level if broad change is to come about.[6]

Where unionization has been successful, workers have gained some notable benefits. At MXP5, only a minority of workers have joined the mainstream unions, like CGIL and CISL, that have a presence in the warehouse.[7] But all have obtained more control over scheduling and salary increases for night shifts. Ultimately, union presence has been successful at establishing a formal relationship with the company and curbing some of the most hideous forms of exploitation. At least for the core group of full-time workers. While overtime is no longer mandatory for MXP5's full-time employees, temp workers are still unable to say no, as flexible schedules are built into their precarious contracts, although in the words of union organizer Andrea:

> In the last couple of years, even workers hired by staffing agencies have noticed that it has become increasingly difficult to get a full-time job simply by saying yes to any managerial demand. Most have been employed at MXP5 four or five times, but always just for a few months. They know it is a precarious job and thus started refusing overtime too.

Disillusionment toward the company's promises have brought more temp workers to realize they have to fight to gain improvements. For instance, the unions have appealed to the labor board to reduce Amazon's use of outsourced workers hired by staffing agencies—CGIL alone represented dozens of these temp workers in the dispute—but eventually the company was not forced to hire them through a regular full-time contract. Amazon knows that a stable workforce is difficult to dominate. Winning higher labor protections and thus constraining the company's system of planned obsolescence for workers would curb its power over workers.

But much more is needed to substantially transform labor conditions for the better. If the warehouse is the new factory, then perhaps old tactics of the labor movement can be renewed, like slowdowns or disruption of the labor process.[8] North American organizers have put renewed attention on some of these traditional industrial tactics. Calls for volunteers to salt Amazon, that is to get militants hired with the specific goal of analyzing the political conditions on the ground and mobilizing the workforce, have appeared in both the United States and Canada.[9] Worker-led collectives branded Amazonians United are agitating in US warehouses, using radical democratic decision-making models and an array of tactics ranging from walkouts to petitions. Often the activists who take up a warehouse job to agitate also aim at "mapping out the organic leaders emerging spontaneously in the warehouse and trying to show how much Amazon betrayed them," as I was told by a salt working to organize Amazonians in a fulfillment center on the US West Coast. That is what MXP5 workers have been doing too: "we are raising a new generation of young workers who are learning, who are interested […] Now for the first time they know there is a group of people that do not give in," Lisa told me in 2021. Although many warehouse employees have no direct experience of labor conflicts or perceive unions as remnants from a different era, the media attention, the organizing on the ground, and the early conquests obtained through these tactics are facilitating the politicization of more workers.

Yet reaching out to the precarious is difficult. If mainstream unions are currently organizing dozens of MXP5 full-time employees, they struggle to include the other half of the workforce. Lisa admitted that her union has "a very low incidence among the hyper-precarious" seasonal and temp workers. They do not have formal union representation in the warehouse. Yet, the potential is there. That may not be that different from the early 1960s FIAT plants analyzed by Romano Alquati, who identified the political role of "new subjects" who arrived as the result of major waves of internal migration from the impoverished South to the industrialized north of Italy. In his telling, unions found it impossible to communicate with this new mass of workers hired to staff the production lines. And yet, Alquati foresaw their political promise, which was to explode a few years later at FIAT and beyond, when the industrial working class occupied the front stage of revolutionary politics in 1960s and 1970s Italy. The story resembles that of American trade unions striving to organize migrant labor in the booming days of early 20th-century industrial capitalism.[10] At Amazon, and especially in a country like Italy, this would mean overcoming the challenges generated by the internal diversity of the workforce, that is, move from the predominantly white working-class workers that are members of mainstream unions, to the migrant workers who represent a major chunk of the temporary workforce.

Traditional unionism has a hard time organizing temp workers. The ethnic composition of this sector of the workforce is certainly an issue, but only when contrasted to the homogeneity of some unions. At global meetings of mainstream unions in Berlin and Dublin, I walked into rooms with predictable red walls and chairs, and faced equally predictable rows of white male speakers—me included. This may soon change. Most warehouses around Piacenza are organized by SI Cobas, the independent union that leads labor in the local logistics industry. The backbone of some of this union's victorious mobilizations have been migrant workers—mostly from the Maghreb—at IKEA or GLS, and young precarious women in successful strikes at Swedish corporation H&M's warehouse. Their actions, based on sit-ins and blockades, were initially inspired by Arab Spring uprisings, connecting labor organizing with the style and tactics of social movements.

In some Amazon fulfillment centers, similar dynamics are unfolding. At BHM1 in Alabama, an FC whose workforce composition is more than 80% Black, workers have depicted the unionization fight conducted by the Retail, Wholesale and Department Store Union (RWDSU) as a fight against Amazon's exploitation of Black workers. Like SI Cobas members from the Maghreb, BHM1 workers were inspired and galvanized by another movement: Black Lives Matter. At the MSP1 fulfillment center in Shakopee, near Minneapolis, struggles against the corporation have been led by the Awood Center, a worker center that organizes East African Amazonians—in Somali, Awood means "solidarity." Led by women in hijab, they have staged walkouts and protests against the warehouse's inhumane productivity standards and overwhelmingly white management.

Should it expand across the entire Amazon workforce, a broader and more diverse participation has the potential to connect struggles across a variety of terrains.[11] With a wider recomposition that includes casualized and migrant workers, with their demands, communities, and political styles, Amazon mobilizations could prove explosive for the future of the logistics industry and perhaps digital capitalism as a whole.

SLACKING OFF AND QUITTING

In the meantime, workers, especially temp workers, need to survive the warehouse and its daily grind. The surveillance system that envelopes work at Amazon means everything associates do is monitored, tracked, counted, and turned into information made readily available to management. Nevertheless, many resort to small individual acts of resistance and sabotage: be it slacking off, deploying tricks to speed up picking and thus maximize their break time, misplacing items on the shelves, or stealing. Remember the MXP5 stower who would grab a comic book, read it, and then place it somewhere it can never be found by Amazon and its algorithms?

Other individual tactics are much more visible. At the beginning of the COVID-19 pandemic, in March 2020, absenteeism became common at MXP5—not just to cope but literally to survive. The province of Piacenza, where MXP5 is located, was a major early hotspot, but social distancing was impossible to maintain in the warehouse, workers were scared and things still uncertain, and the company was not providing appropriate personal protection equipment. So, many MXP5 associates started using sick leave en masse. While the tactic was certainly not sustainable, many saw it as a last resort. Like other workers, Lisa confirmed that "Many were just not showing up. You would get to work and wonder 'where is that guy, where is that other colleague,' but management would not tell you whether they were sick or quarantined, or whether they were just skipping work." According to some estimates, up to 30% of the workforce was not clocking in for work.[12] Anthropologist James Scott called these everyday forms of resistance that blossom even under the most oppressive conditions "weapons of the weak."[13]

The most definitive individual expression of refusal is the act of quitting. Those workers who have employment alternatives, who have become terminally fed up with the Amazon pace or systems of sur-veillance, or feel they can no longer cope with management by stress, leave the warehouse in search of better opportunities elsewhere: a sort of bottom-up flexibility. Many of the workers I talked to while writing this book have now left MXP5. Those who did so voluntarily did not regret leaving.

Quitting has different meanings and effects that follow lines of privilege, as only some can quit without jeopardizing their future economic security or employability. At the height of the first wave of the COVID-19 pandemic in April 2020, Tim Bray, then Vice President of Amazon Web Services, resigned. He did it because: "remaining an Amazon VP would have meant [...] signing off on actions I despised," since "Amazon treats the humans in the warehouses as fungible units of pick-and-pack potential."[14] It has become relatively common for executives in Big Tech companies to quit and go public about it, particularly in the wake of a recent decline of the myth of tech's contribution to the greater good of humanity. Media theorists Tero Karppi and David Nieborg have analyzed similar "corporate abdications" at Facebook, suggesting they be understood as part of a broader "epochal wave of techno-dystopianism" that even company executives feel guilty about.[15] Of course, the exits of executives and rank-and-file workers cannot be equated. For the executives, leaving is not only a moral gesture but also a form of redemption: they leave as an act of self-improvement. Michel Foucault would have interpreted these audience-oriented actions as performances of redemption and purification.[16]

On the contrary, not many warehouse associates have this luxury, nor do they need to be purified in the first place: they shoulder the effects of, not the responsibility for, techno-dystopianism. For many, quitting is simply a way out of physical and mental breakdown. And yet, the ritual of publicizing a choice to quit is performed by warehouse workers too. The "Why I quit Amazon" YouTube video has become a genre in itself, along with countless social media and blog posts. Amazon (ex)associates produce and share these videos to resist the internalization of responsibility and to reject the blame for not being able to keep up with the pace of Amazon work. They need to do this because at Amazon workers are taught to be responsible for their own failure. On the warehouse floor, this means that injuries or the inability to meet target quotas are attributed to the worker's incorrect work practices ("Why did you drink so much water? Why didn't you get enough sleep between back-to-back 12-hour shifts?"). By publicly refusing to bear the brunt of failure, workers who have left Amazon aim to share their experiences with breakdown and burnout, and together develop a mutual understanding of what went wrong.[17] These confessionals are part of a general need to create a community online. Many use online spaces such as Reddit or other forums and social media platforms to find each other and share their work experiences. In the absence of spaces of socialization that escape the pervasive surveillance system inside the warehouse, these forums are the de facto break rooms where workers meet and agitate.[18] Both video confessionals and forums may help build support networks of ex-workers that allow them to cope with the psychological effects of failure and breakdown.

Often, though, quitting leads to more atomization.Even when it is beneficial for the individual worker, this individual form of refusal does not bring about collective power. Actually, high turnover rates represent a concrete barrier to organizing, as many disgruntled workers quit instead of fighting within Amazon. As I was told by the West coast organizer, "the moment they are agitated enough to get involved, they are also agitated enough to quit." And then, quitting to go where? Alternatives to Amazon are harder and harder to find. In some areas, Amazon has become a major employer, destroying jobs in retail and catalyzing a restructuring of the labor market. Workers in Piacenza, like those in many similar areas where Amazon has a fulfillment center, may have more opportunities, but many of those other opportunities will be for a very similar job, just in another company's warehouse, be it IKEA, TNT, or Zalando. Finally, quitting may play into Amazon's hands, since the company does not exactly shy away from worker turnover. Rather, it treats it as a core feature of its business model, going so far as to engender planned employee obsolescence. The company is perfectly fine with people leaving in droves, so long as a reserve army of new workers is available to replace them. In the long run, it is planning to introduce automation that reduces its dependence on human labor—although it may never be able to do away with it. Unless it is politicized and generalized, refusal of work will remain a self-preserving strategy but won't make a dent in Amazon's power.

SUBVERTING AMAZON

What organized workers seek to do is fight simultaneously within, against, and beyond Amazon. To do so, workers subvert the company's very characteristics, rather than deserting it.

It is both a material and symbolic struggle. Questioning the company's techno-idyllic myth by unboxing the reality behind the smiling arrow is crucial. It won't be easy: the myth of progress and emancipation built by Amazon and the managerial techniques it uses inside the warehouse to capture workers' consent are often successful. Many like working at Amazon and enjoy the FCs' culture of fun. On workplace review platforms like Glassdoor, Amazon receives high grades from many of its workers. But as put by a union organizer who had been working to unionize MXP5, it's important to "outflank it, hit where it's weak, that is the smile, the facade. We know it's fake." For example, a delegation of workers from Piacenza traveled to Berlin in April 2018, joining Amazonians from all over Europe to protest Jeff Bezos as he was receiving an award given yearly to people who have "exceptional talent for innovation [...] and also face their social responsibility."[19] The bar for social responsibility must have been quite low when Bezos was selected, workers reminded the award committee. Workers have scratched the myth also by repurposing Amazon's own slogans, such as *customer obsession*. Worker collective Amazon Employees for Climate Justice did just that when they crashed the 2019 shareholder meeting to ask the company to deal with its environmental impact. At the meeting, they positioned themselves as model Amazon employees, telling the top brass that by contributing to destroying the planet through the company's massive carbon footprint they were not showing customer obsession.[20]

Most importantly, Amazon workers have the potential to undermine and disrupt operations, and thus leave the company's promise of the quick realization of consumers' desires unfulfilled. Indeed, the flexible model built by Amazon is vulnerable. It relies on a few major bursts of activity, and those moments can be targeted to maximize the effects of a strike or walkout. MXP5 has gone on strike around Black Friday, and even before the first strike in 2017, workers had started declaring a state of agitation during peaks to safeguard workers willing to refuse the imposition of overtime exactly when Amazon needs it the most.[21] In other countries too, worker actions have been organized on Prime Day or Cyber Monday.

Amazon's fulfillment system is organized around bottlenecks that are crucial for the circulation of commodities: nothing can be delivered unless it is retrieved, packed, and shipped from the fulfillment center. This can be put to tactical advantage to break down Amazon's supply chain.[22] Yet if the ambition is to have a real effect on the company's bottom line, blocking individual fulfillment centers is not enough: Amazon has built redundant networks of warehouses that allow it to shift orders and avoid major disruptions if one stops working. This problem is readily acknowledged by workers, but they are also finding ways around it. For instance, in 2018, one of MXP5's union delegates stressed how

> strikes are very difficult to organize and manage. You need to unionize as much as possible. In MXP5 we did a good job, but what about Rome and Vercelli? If you don't stop Rome they can shift orders, and there you go, it's over; maybe you create some difficulties but not such a big mess. We need time.

Furthermore, thanks to the flexible workforce it employs in its facilities, Amazon has the ability to tap into its masses of temp workers to make up for the labor shortage caused by labor action. So, a plurality of strategic locations must be contested at the same time, or Amazon will find ways to circumvent a strike by moving orders around its network of FCs, or by relying on precarious workers who are more difficult to agitate and mobilize.

To confront Amazon's resilient logistical system, what is needed is a counterlogistics of struggles[23]—a networked organization that has the ability to rise up tactically to overcome the organizational and technological obstacles put in place by the corporation. To lay siege to Amazon.

The first such attempt to besiege Amazon in this way took place in Italy on March 22, 2021. It was a national strike organized by mainstream unions and involving Amazon's entire distribution chain. In particular, it demanded economic bonuses for minimum-wage workers, and a limitation to work rhythms, especially for delivery couriers. That day, many warehouse associates from MXP5 and other fulfillment centers stopped working, but the strike was designed to disrupt Amazon's entire distribution structure, and in fact the main effects were generated by stoppages at the seven small last-mile distribution centers that surround downtown Milan, as well as in other cities in the rest of the country. If warehouses and distribution centers store and manage commodities, it is the drivers and couriers of the companies contracted by Amazon that deliver the packages to the consumer. They also went on strike—among others, hundreds of couriers who for a year had been protesting the long shifts and intense rhythms of work imposed by Amazon. Unions estimated that in Lombardy alone, 250,000 orders were not delivered that day. Even the national call center in Sardinia was affected. Another union organizer I spoke with, Andrea, noted a further difference between the 2021 and the 2017 strikes. Not only the initial handful of militant workers had grown to quite a sizable chunk of MXP5's workforce, but also unions had finally managed to establish a presence in other Italian fulfillment centers too: "now we talk to each other, we can move the strike to Vercelli and Rome, and Amazon can no longer shift orders through those warehouses." The strike demonstrated how thanks to this solidarity, worker action can synchronize, flex, and scale-up, exactly like Amazon's fulfillment processes.

An alliance across Amazon's distribution chain is crucial, but what about moving beyond it to embrace a broader process of recomposition that unites workers in different positions in the company's global division of labor? This coalition model is blossoming but may need to become even broader than the one mobilized in Italy in March 2021. For instance, forms of inter-class solidarity between warehouse and delivery workers and engineers are already in place. Amazon engineers and tech workers have been part of campaigns to denounce the conditions of work faced by warehouse associates. Compared to the warehouse, the flashy downtown Seattle offices where engineers and other white-collar employees work are just another planet. They employ a predominantly white workforce, pay a higher salary, and provide better benefits and a better work environment. Yet what they share with warehouse associates are the same concerns for the despotic nature of Amazon's management, as well as the human and environmental impact of its e-commerce operations. Organizing across entire communities has also helped warehouse workers by positioning their struggles within a broader fight against sexism, institutional racism, and austerity. Local environmental movements are opposing the construction of new fulfillment centers, protesting the impactful land use and atmospheric pollution they force upon communities.[24] Global campaigns bring together broad coalitions of NGOs, worker centers, and social movements.[25]

Amazon workers are not alone in this titanic fight.[26] They know that the battle for the warehouse cannot be waged in isolation from the other struggles against global digital capitalism. Workers from Toronto to Jakarta have shown that attempts at individualizing and separating workers from each other do not prevent the emergence of collective structures of solidarity. The conditions of possibility for organizing the supposedly unorganizable, algorithmically managed, individualized laborers of digital capitalism are already there, and are being seized by workers across the globe.[27] Worker uprisings are now being staged across many emerging forms of work underpinned by digital technology, from engineers organizing in Google campuses, to transnational alliances among food couriers working for apps such as Foodora or Deliveroo and drivers working for car-sharing industry companies like Uber. These struggles have proven that bottom-up worker movements can imagine new forms of self-organization that are adequate to today's technological and political challenges and do away with the limitations and delays of traditional union politics.[28]

CODA

Amazon is still in the midst of an expansion phase. It is currently building new warehouses at an unprecedented rate, including in countries where it has only recently begun operating, like India. It is also going through an unprecedented cycle of accumulation of both capital and power. The millions of people Amazon puts to work contribute to this expansion, but precisely because workers keep it going they also have the power to undermine it. This is also why workers' knowledge and experiences are so central. Countering Amazon's corporate power will involve a better understanding of how this digital factory both organizes work and rearranges social and power relations to ensure its relentless accumulation.

And yet, Amazon's e-commerce empire is fragile. It has high operational costs and low profit margins, especially compared to other services it offers, like its cloud computing AWS. The increasingly global worker unrest it faces is a fundamental threat to its ability to sustain its system of exploitation. Should they materialize, calls for breaking up Amazon through antitrust laws would affect the corporation's ability to build monopolies, but also the integration of computing, e-commerce and surveillance its power over workers is based upon. Negotiating or refusing the introduction of new automation, or demanding the abolition of workplace surveillance, workers could steer the course of Amazon's trajectory in the innovation of inequity. But none of that may be enough. Perhaps we will need to put an end to digital capitalism corporations altogether as the only way to liberate technology from their grip.[29] Some have suggested that Amazon's ability to plan global production and distribution chains can and should be turned toward democratic goals—socialize it and give all power to its workers.[30]

What's certain is that a failure to imagine new social forms and new ways to work independently of capitalist relations of production is also what gives Amazon the power to make history the way it wants it. That must also be subverted. To reclaim the future. To leave the warehouse.

Notes

1 RELENTLESS

1. Warde, A. (2015). The sociology of consumption: Its recent development. *Annual Review of Sociology, 41,* 117–134. See also Graeber, D. (2011). Consumption. *Current Anthropology, 52*(4), 489–502.
2. Marx, K. (1867, 1976). *Capital* (Vol. 1). Penguin, p. 125. Marx discussed circulation in the second volume of Capital.
3. Ciccarelli, R. (2018). *Forza lavoro: Il lato oscuro della rivoluzione digitale.* DeriveApprodi, pp. 10–12.
4. Quoted in Stone, B. (2013). *The everything store: Jeff Bezos and the age of Amazon.* Little, Brown and Company, p. 286.
5. For an analysis of Amazon as a new form of capitalism see Alimahomed-Wilson, J., and Reese, E. (2020). *The cost of free shipping: Amazon in the global economy.* Pluto Press.
6. Bergvall-Kåreborn, B., and Howcroft, D. (2014). Amazon Mechanical Turk and the commodification of labour. *New Technology, Work and Employment, 29*(3), 213–223.
7. See Cowen, D. (2014). *The deadly life of logistics: Mapping violence in global trade.* University of Minnesota Press; and Orenstein, D. (2019). *Out of stock: The warehouse in the history of capitalism.* University of Chicago Press.
8. Amnesty International. (2020). *Public statement: It is time for Amazon to respect workers' right to unionize.* Retrieved from www.amnesty.org.
9. Bezos, J. (1998). *1997 letter to shareholders.* Retrieved from www.aboutamazon.com.
10. Charam, R., and Yang, J. (2019). *The Amazon management system.* IdeaPress Publishing. See also Galloway, S. (2017). *The four: The hidden DNA of Amazon, Apple, Facebook, and Google.* Penguin; and Dumaine, B. (2020). *Bezonomics: How Amazon is changing our lives and what the world's best companies are learning from it.* Scribner.
11. Stone, B. (2013). *The everything store: Jeff Bezos and the age of Amazon.* Little, Brown and Company, pp. 176–177.
12. Kantor, J., and Streitfield, D. (2015, August 15). Inside Amazon: Wrestling big ideas in a bruising workplace. *The New York Times.*
13. Klein, N. (2007). *Shock doctrine.* Knopf Canada.
14. Goodwin, H. (2020, December 10). Jeff Bezos could give all Amazon workers $105,000 and still be as rich as pre-Covid. *The London Economic.*

15. Ligman, K. (2018). *You are Jeff Bezos*. Retrieved from https://direkris.itch. io/you-are-jeff-bezos.
16. The quote is taken from Allen Mandelbaum's English translation. See Alighieri, D. (1982). *Inferno*. Random House, canto I, 98–99.
17. Braudel, F. (1995). *The Mediterranean and the Mediterranean world in the age of Philip II*. University of California Press, p. 379.
18. Arrighi, G. (1994). *The long twentieth century: Money, power, and the origins of our times*. Verso.
19. Massimo, F. (2020, April 24). Piacenza, il virus e il container. *Il Mulino*.
20. Heel, P. (2018). *Hinterland: America's new landscape of class and conflict*. Reaktion Books.
21. Apicella, S. (2020). Rough terrains: Wages as mobilizing factor in German and Italian Amazon distribution centers. *Sozial Geschichte Online, 27*, 1–14.
22. Rossiter, N. (2017). *Software, infrastructure, labor: A media theory of logistical nightmares*. Routledge, p. 5.
23. An API, or application programming interface, is a software interface that connects computers or computer programs. The quote is taken from Bezos, J. (2007). *2006 letter to shareholders*. Retrieved from www. aboutamazon.com.
24. Jarrett, K. (2003). Labour of love: An archaeology of affect as power in e-commerce. *Journal of Sociology, 39*(4), 335–351.
25. Crawford, K., and Joler, V. (2018). *Anatomy of an AI system*. Retrieved from http://anatomyof.ai.
26. Huws, U. (2014). *Labor in the global digital economy: The cybertariat comes of age*. Monthly Review Press.
27. ChainCrew (2002). *ChainWorkers. Lavorare nelle cattedrali del consumo*. DeriveApprodi. On non-places see Augé, M. (1992). *Non-lieux. Introduction à une anthropologie de la surmodernité*. Seuil.
28. Biagioli, M. (2014). *Celebrating garages, mythifying Silicon Valley* [Conference presentation]. European University at St. Petersburg. On the garage as a site of innovation see also Erlanger, O., and Ortega Govela, L. (2018). *Garage*. MIT Press. On its gendered nature see Jen, C. (2015). Do-it-yourself biology, garage biology, and kitchen science. In: Wienroth, M., and Rodrigues, E. (eds.), *Knowing new biotechnologies: Social aspects of technological convergence*. Routledge, pp. 125–141.
29. This was described as "Californian ideology" by British cultural critics Barbrook, R., and Cameron, A. (1996). The Californian ideology. *Science as Culture, 6*(1), 44–72.
30. For an analysis of letters from the founders of Google, Groupon, and Facebook, see Dror, Y. (2015). "We are not here for the money": Founders' manifestos. *New Media & Society, 17*(4), 540–555. Or for Twitter and Yelp, see Nam, S. (2020). Cognitive capitalism, free labor, and financial communication: A critical discourse analysis of social media IPO registration statements. *Information, Communication & Society, 23*(3), 420–436.

31. Nakamura, L. (2014). Indigenous circuits: Navajo women and the racialization of early electronic manufacture. *American Quarterly*, 66(4), 919–941.

32. Quoted in Steinberg, J. (2012, October 18). Amazon creates 700 jobs in San Bernardino with new distribution center. *The Sun*; and in Skebba, J. (2019, July 22). It's official: Amazon is coming to Rossford. *The Blade*.

33. Jones, J., and Zipperer, B. (2018). *Unfulfilled promises: Amazon fulfillment centers do not generate broad-based employment growth.* Economic Policy Institute.

34. Anonymous. (2018, January 20). Unfulfillment centres: What Amazon does to wages. *The Economist*.

35. Huws, U. (2014). *Labor in the global digital economy: The cybertariat comes of age.* Monthly Review Press, p. 39.

36. Apicella, S. (2020). Rough terrains: Wages as mobilizing factor in German and Italian Amazon distribution centers. *Sozial Geschichte Online*, 27, 1–14.

37. Adecco. (2021). *MOG: Monte ore garantito.* Retrieved from www.adecco.it.

38. See Amazon (2020). *Our workforce data.* Retrieved from www.aboutamazon.com.

39. For instance, in Veneto. See Ferro, E. (2021, January 10). Licenziato da Amazon il magazziniere costretto a vivere in camper. *La Repubblica*.

40. The League of Cooperatives, or Legacoop, is a national organization founded in the late 19th century. It federates thousands of cooperatives, ranging from small worker-led organizations to a major supermarket chain. It is particularly powerful in Emilia-Romagna. Historically connected to the communist left, it is the "red" federation as opposed to the "white," or catholic, Confcooperative.

41. Altenried, M. (2020). The platform as factory: Crowdwork and the hidden labour behind Artificial Intelligence. *Capital & Class*, 44(2), 145–158.

42. Early workerists were a pre-1968 loose group of Italian political activists and intellectuals organized around militant publications such as *Quaderni rossi* and *Classe operaia*. Many had left the Italian Communist Party to study the new forms of industrial capitalism that were emerging in the most advanced areas of Northern Italy, and the struggles through which workers resisted them. On this story see Wright, S. (2017). *Storming heaven: Class composition and struggle in Italian Autonomist Marxism.* Pluto Press.

43. Benjamin, R. (2016). Innovating inequity: If race is a technology, postracialism is the Genius Bar. *Ethnic and Racial Studies*, 39(13), 2227–2234.

2 WORK HARD

1. As reported by a former Amazon executive: see Kantor, J., Weise, K., and Ashford, G. (2021, June 15). The Amazon that customers don't see. *The New York Times*.

2. On digital Taylorism at Amazon see Massimo, F. (2019). Spettri del Taylorismo. Lavoro e organizzazione nei centri logistici di Amazon. *Quaderni di Ricerca Sociale, 3,* 85–102. For a more general theorization see Altenried, M. (2020). The platform as factory: Crowdwork and the hidden labour behind Artificial Intelligence. *Capital & Class, 44*(2), 145–158.

3. Walker, T. (2020). Alexa, are you a feminist? Virtual assistants doing gender and what that means for the world. *iJournal, 6*(1), 1–16.

4. On social media see Postigo, H. (2016). The socio-technical architecture of digital labor: Converting play into YouTube money. *New Media & Society 18*(2), 332–349, and Cohen, N. (2018). At work in the digital newsroom. *Digital Journalism, 7*(5), 571–591. On ride-hailing see Rosenblat, A., and Stark, L. (2016). Algorithmic labor and information asymmetries: A case study of Uber's drivers. *International Journal of Communication, 10,* 3758–3784; or Chen, J. (2017). Thrown under the bus and outrunning it! The logic of Didi and taxi drivers' labour and activism in the on-demand economy. *New Media & Society, 20*(8), 2691–2711. On crowdwork see Bergvall-Kåreborn, B., and Howcroft, D. (2014). Amazon Mechanical Turk and the commodification of labour. *New Technology, Work and Employment, 29*(3), 213–223; or Lee, M.K., Kusbit, D., Metsky, E., and Dabbish, L. (2015). Working with machines: The impact of algorithmic and data-driven management on human workers. *Proceedings of the 33rd Annual ACM Conference on Human Factors in Computing Systems,* USA, pp. 1603–1612.

5. Aneesh, A. (2009). Global labor: Algocratic modes of organization. *Sociological Theory, 27*(4), 347–370; see also Danaher, J. (2015). The threat of algocracy: Reality, resistance and accommodation. *Philosophy of Technology, 29,* 245–268.

6. For an analysis of the methods that one can use to study proprietary algorithms see Burrell, J. (2016). How the machine "thinks": Understanding opacity in machine learning algorithms. *Big Data & Society, 3*(1); Kitchin, R. (2016). Thinking critically about and researching algorithms. *Information, Communication & Society, 20*(1), 14–2; or Seaver, N. (2017). Algorithms as culture: Some tactics for the ethnography of algorithmic systems. *Big Data & Society, 4*(2), 1–12.

7. Alquati, R. (1975). *Sulla FIAT e altri scritti.* Feltrinelli, pp. 114–117.

8. On the role of workers in training and maintaining automated systems see Casilli, A. (2018). *En attendant les robots: Enquête sur le travail du clic.* Seuil.

9. See for example Muralidhara, G.V., and Vijai, P. (2016). *Inside Amazon: Chaotic storage system.* IBS Center for Management Research. Retrieved from www.thecasecentre.org; or Hausman, W.H., Schwarz, L.B., and Graves, S.C. (1976). Optimal storage assignment in automatic warehousing systems. *Management science, 22*(6), 629–638.

10. Agre, P.E. (1994). Surveillance and capture: Two models of privacy. *The Information Society, 10*(2), 101–127.

11. On data capture in logistics see Rossiter, N. (2017). *Software, infrastructure, labor: A media theory of logistical nightmares.* Routledge, pp. 4–5.

12. Weidinger, F., and Boysen, N. (2018). Scattered storage: How to distribute stock keeping units all around a mixed-shelves warehouse. *Transportation Science, 52*(6), p. 1412.

13. As noted by Jesse LeCavalier in his book on the organization of inventory in another giant corporation: Walmart. LeCavalier described how Walmart workers extend software systems—taking up tasks that computers cannot perform, while at the same time generating information that can be utilized by computers to manage other crucial operations. See LeCavalier, J. (2016). *The rule of logistics: Walmart and the architecture of fulfillment.* University of Minnesota Press, p. 153.

14. On this process see Danaher, J. (2015). The threat of algocracy: Reality, resistance and accommodation. *Philosophy of Technology, 29,* 245–268.

15. This is the subject of an entire field of research, but on warehousing specifically see Gertler, M.S. (2003). Tacit knowledge and the economic geography of context, or the undefinable tacitness of being (there). *Journal of Economic Geography, 3*(1), 75–99.

16. See Chen, J. (2017). Thrown under the bus and outrunning it! The logic of Didi and taxi drivers' labour and activism in the on-demand economy. *New Media & Society, 20*(8), 2691–2711; and Cant, C. (2019). *Riding for Deliveroo: Resistance in the new economy.* Polity.

17. For a history of the warehouse within this trend see Orenstein, D. (2019). *Out of stock: The warehouse in the history of capitalism.* University of Chicago Press. For an analysis of logistics as a lens to analyze contemporary capitalism see Benvegnù, C., Cuppini, N., Frapporti, M., Milesi, F., and Pirone, M. (2019). Logistical gazes: introduction to a special issue of *Work Organisation, Labour and Globalisation. Work Organisation, Labour & Globalisation, 13*(1), 9–14.

18. "Degradation of labor" is a famous definition coined by sociologist Harry Braverman in his book about factory work: Braverman, H. (1974). *Labour and monopoly capital: The degradation of work in the twentieth century.* Monthly Review Press. For accounts written by FC workers themselves see for instance (among many others) Amazon workers (2018). Stop treating us like dogs! Workers organizing resistance at Amazon in Poland. In: Alimahomed-Wilson, J. and Ness, I. (eds.), *Choke points: Logistics workers disrupting the global supply chain.* Pluto Press, pp. 96–109; or Anonymous. (2018, November 21). Our new column from inside Amazon: They treat us as disposable. *The Guardian.*

19. For a historical account of the tyranny of clocks see Thompson, E.P. (1967). Time, work-discipline, and industrial capitalism. *Past & Present, 38,* 56–97.

20. Brar, A., Daniel, M., and Sra, G. (2020, December 30). "I am scared to take a day off whether sick or not." The voiceless warehouse workers in Peel and how COVID-19 has silenced them even more. *The Toronto Star.*

21. Strategic Organizing Center. (2021, May). *Primed for speed: Amazon's epidemic of workplace injuries*. Retrieved from www.thesoc.org.
22. Evans, W. (2020, September 29). How Amazon hid its safety crisis. *Reveal*.
23. Tung, I., and Berkowitz, D. (2020, March 6). *Amazon's disposable workers: High injury and turnover rates at fulfillment centers in California*. National Employment Law Project Data Brief; for a summary of the report see Evans, W. (2020, September 29). How Amazon hid its safety crisis. *Reveal*.
24. Bezos, J. (2021). *2020 letter to shareholders*. Retrieved from www.aboutamazon.com.
25. Evans, W. (2020, September 29). How Amazon hid its safety crisis. *Reveal*.
26. Huws, U. (2014). *Labor in the global digital economy: The cybertariat comes of age*. Monthly Review Press.
27. Marx, K. (1867, 1976). *Capital* (Vol. 1). Penguin, p. 526.
28. Braverman, H. (1974). *Labour and monopoly capital: The degradation of work in the twentieth century*. Monthly Review Press, p. 82.
29. Panzieri, R. (1961). Sull'uso capitalistico delle macchine nel neocapitalismo. *Quaderni rossi*, *1*, p. 74.
30. Zuboff, S. (2019). *The age of surveillance capitalism*. PublicAffairs.
31. Couldry, N., and Mejias, U.A. (2019). *The costs of connection: How data is colonizing human life and appropriating it for capitalism*. Stanford University Press; see also Sadowski, J. (2019). When data is capital: Datafication, accumulation, and extraction. *Big Data & Society*, 1–12.
32. For an early analysis of the commodification of consumer surveillance see Cohen, N.S. (2008). The valorization of surveillance: Towards a political economy of Facebook. *Democratic Communiqué*, *22*(1), 5–22.
33. Tronti, M. (1966, 1980). The strategy of refusal. *Semiotext(e)*, *3*(3), 29–32.
34. Wajcman, J. (2014). *Pressed for time*. University of Chicago Press.
35. Beniger, J. (1986). *The control revolution: Technological and economic origins of the information society*. Harvard University Press.

3 HAVE FUN

1. As in many other sectors of the digital economy. See Huws, U. (2016, June 1). Logged in. *Jacobin*.
2. Amazon (2021, May 17). *From body mechanics to mindfulness, Amazon launches employee-designed health and safety program called WorkingWell across U.S. operations* [Press release]. Retrieved from www.aboutamazon.com.
3. See for instance Gregg, M. (2018). *Counterproductive: Time management in the knowledge economy*. Duke University Press.
4. Quoted in Hahn, J. (2019, November 27). I went on a propaganda tour of an Amazon warehouse. *VICE*.

5. Fleming, P. (2005). Workers' playtime? Boundaries and cynicism in a "culture of fun" program. *The Journal of Applied Behavioral Science, 41*(3), 285–303.

6. As common in industries from social media to green economies. On social media see Duffy, B.E. and Schwartz, B. (2018). Digital "women's work?": Job recruitment ads and the feminization of social media employment. *New Media & Society, 20*(8), 2972–2989; on green economies see Castellini, V. (2019). Environmentalism put to work: Ideologies of green recruitment in Toronto. *Geoforum, 104*, 63–70.

7. Amazon (2021). *Employee engagement*. Retrieved from www.about amazon.com.

8. Anonymous (2017). Working for Amazon: Better than sex, worse than hell (Part 1). *Naked Capitalism*.

9. Amazon (2021). *Employee engagement*. Retrieved from www.about amazon.com.

10. Amazon (2021). *Employee engagement*. Retrieved from www.about amazon.com.

11. Mollick, E. R. and Rothbard, N. (2014). Mandatory fun: Consent, gamification and the impact of games at work. *The Wharton School research paper series*.

12. On call centers see Brophy, E. (2017). *Language put to work: The making of the global call centre workforce*. Palgrave Macmillan; and Woodcock, J. (2017). *Working the phones: Control and resistance in call centres*. Pluto Press. On gamification in the gig economy see Mason, S. (2018, November 20). High score, low pay: Why the gig economy loves gamification. *The Guardian*.

13. Burawoy, M. (1979). *Manufacturing consent. Changes in the labor process under monopoly capitalism*. University of Chicago Press.

14. Dow Schüll, N. (2012). *Addiction by design: Machine gambling in Las Vegas*. Princeton University Press.

15. Rosenblat, A. and Stark, L. (2016). Algorithmic labor and information asymmetries: A case study of Uber's drivers. *International Journal of Communication, 10*, 3758–3784.

16. Woodcock, J., and Johnson, M. (2018). Gamification: What it is, and how to fight it. *The Sociological Review, 66*(3), 542–558; see also Fizek S., Fuchs, M., Ruffino, P., and Schrape, N. (eds.). (2014). *Rethinking gamification*. Meson Press.

17. Han, B. (2017). *Psychopolitics: Neoliberalism and new technologies of power*. Verso.

18. Open Markets Institute. (2020, August 31). *Eyes everywhere: Amazon's surveillance infrastructure and revitalizing worker power*.

19. For a first-hand account of this phenomenon in Canada see Amazon Workers Collective (2020, October 28). Workers of the world: Salt at Amazon! *Briarpatch*.

20. Gurley, L.K. (2020, October 7). Pregnant Amazon employees speak out about nightmare at Oklahoma warehouse. *Vice*.

21. Dastin, J. and Hu, H. (2020, April 18). Exclusive: Amazon deploys thermal cameras at warehouses to scan for fevers faster. *Reuters*.

22. Schreiber, E. (2020, December 11). Amazon Web Services offers companies new tools for spying on workers. *World Socialist Web Site*.

23. Peterson, H. (2020, April 20). Whole Foods tracks unionization risk with heat map. *Business Insider*.

24. Yeturu, K., and Huddleston, H.L. (2019). Image creation using geo-fence data. US patent 10313638.

25. American Civil Liberties Union (2020, June 10). *ACLU statement on Amazon face recognition moratorium*. Retrieved from www.aclu.org

26. Ali, H. (2020, September 25). Amazon's surveillance system is a global risk to people of color. *Medium*.

27. Browne, S. (2015). *Dark matters. On the surveillance of blackness*. Duke University Press.

28. The ad is still visible at https://web.archive.org/web/20200901125940/ https://www.amazon.jobs/en/jobs/1026060/intelligence-analyst

29. Gurley, L.J. (2020, November 23). Secret Amazon reports expose the company's surveillance of labor and environmental groups. *VICE*.

30. For a thorough description of Amazon's surveillance system see Delfanti, A., Radovac, L., and Walker, T. (2021). *The Amazon panopticon: A guide for organizers and policymakers*. UNI Global Union.

31. Rosenblat, A. and Stark, L. (2016). Algorithmic labor and information asymmetries: A case study of Uber's drivers. *International Journal of Communication*, *10*, 3758–3784.

32. Kantor, J., Weise, K., and Ashford, G. (2021, June 15). The Amazon that customers don't see. *The New York Times*.

33. Reese, E. and Struna, J. (2018). "Work hard, make history": Oppression and resistance in Inland Southern California's warehouse and distribution industry. In: Alimahomed-Wilson, J. and Ness, I. (eds.), *Choke Points. Logistics Workers Disrupting the Global Supply Chain*. Pluto Press, pp. 81–95. They borrow the idea of management by stress from Parker, M., and Slaughter, J. (1994). Management-by-stress: Management's ideal. In: Parker, M. and Slaughter, J. (eds.), *Working smart: A union guide to participation programs and reengineering*. Labor Notes, pp. 24–38.

34. In the scene, Queen Cersei is humiliated in public as she is forced to walk naked through a crowd.

35. Curcio, A. (2000). Resisting sexism and racism in Italian logistics worker organizing. In: Ovetz, R. (ed.), *Workers' inquiry and global class struggle*. Pluto Press, pp. 90–102.

36. See Salzinger, L. (2003). *Genders in productions: Making workers in Mexico's global factories*. University of California Press. On the role of women's sexualities and sexual harassment in American FCs see Reese, E. (2020). Gender, race, and Amazon warehouse labor in the United States.

In: Alimahomed-Wilson, J., and Reese, E. (eds.) *The cost of free shipping. Amazon in the global economy*. Pluto Press, pp. 102–115.

37. Ghaffari, S. and Del Rey, J. (2020, June 29) The real cost of Amazon. *Vox*.
38. Panzieri, R. (1961). Sull'uso capitalistico delle macchine nel neocapitalismo. *Quaderni rossi*, *1*, p. 63.
39. Ellul, J. (1964). *The technological society*. Knopf, p. 25.
40. Panzieri, R. (1961). Sull'uso capitalistico delle macchine nel neocapitalismo. *Quaderni rossi*, *1*, p. 61.
41. McLuhan, M. (1964). *Understanding media: The extensions of man*. McGraw-Hill.
42. For another use of this metaphor see Gordon, E., and Manosevitch, E. (2011). Augmented deliberation: Merging physical and virtual interaction to engage communities in urban planning. *New Media & Society*, *13*(1), 75–95.
43. Turner, F. (2009). Burning Man at Google: A cultural infrastructure for new media production. *New Media & Society*, *11*(1–2), 73–94.
44. On the role of mindfulness in white collar labor see for instance Guyard, C., and Kaun, A. (2018). Workfulness: Governing the disobedient brain. *Journal of Cultural Economy*, *11*(6), 535–548.
45. ABC News Story Lab. (2019). *The Amazon Race*. Retrieved from https://www.abc.net.au/news/2019-02-27/amazon-warehouse-workers-game-race/10803346?nw=0.
46. Warcraft is a real-time strategy game where players manage workers and resources to build armies and defeat their opponents. The "peon" is an easily killed and easily replaced unit used to chop wood, mine gold, and construct buildings. The quote is taken from Anonymous. (2017). Working for Amazon: Better than sex, worse than hell (Part 1). *Naked Capitalism*.
47. Kunda, G. (2009). *Engineering culture: Control and commitment in a high-tech corporation*. Temple University Press.
48. Fisher, C.D. (2010). Happiness at work. *International Journal of Management Reviews*, *12*(4), 384–412.
49. Bakker, A.B., Schaufeli, W.B., Leiter, M.P., and Taris, T.W. (2008). Work engagement: An emerging concept in occupational health psychology. *Work & Stress*, *22*(3), 187–200.
50. Jackson, N., and Carter, P. (2011). In praise of boredom. *Ephemera: Theory & Politics in Organization*, *11*(4), 388–389.
51. Gregg, M. (2018). *Counterproductive: Time management in the knowledge economy*. Duke University Press.
52. Vincent, J. (2021, June 17). Canon put AI cameras in its Chinese offices that only let smiling workers inside. *The Verge*.
53. Tapscott, D. and Caston, A. (1993). *Paradigm shift: The new promise of information technology*. McGraw Hill.

54. On the relation between happiness and productivity in a culture of fun
see Jackson, N., and Carter, P. (2011). In praise of boredom. *Ephemera:
Theory & Politics in Organization, 11*(4), 387–405.

4 CUSTOMER OBSESSION

1. On just-in-time in logistics see Cowen, D. (2014). *The deadly life of logistics: Mapping violence in global trade.* University of Minnesota Press.
2. Amazon (2021). *Our workplace.* Retrieved from www.aboutamazon.com.
3. Kim, E., and Stewart, A. (2021, May 10). Some Amazon managers say
they "hire to fire" people just to meet the internal turnover goal every
year. *Business Insider.*
4. Quoted in Laucius, J. (2018, August 21). The Amazon effect: Will Ottawa's
new fulfillment centre create "middle class" jobs? *Ottawa Citizen.*
5. Gurley, L. (2021, May 11). Amazon rebrands its brutal "megacycle" shift
to "single cycle." *VICE.*
6. Boushey, H. (2016). *Finding time: The economics of work-life conflict.*
Harvard University Press; McCrate, E. (2012). Flexibility for whom?
Control over work schedule variability in the US. *Feminist Economics,
18*(1), 39–72; Watson, E. and Swanberg, J. (2011). *Rethinking workplace
flexibility for hourly workers: Policy brief.* Georgetown Law, Georgetown
University.
7. See, for instance, Henly, J.R., and Lambert, S.J. (2014). Unpredictable
work timing in retail jobs: Implications for employee work–life conflict.
Ilr Review, 67(3), 986–1016.
8. Chen, J.Y., and Sun, P. (2020). Temporal arbitrage, fragmented rush,
and opportunistic behaviors: The labor politics of time in the platform
economy. *New Media & Society, 22*(9), 1561–1579.
9. Cant, C. (2019). *Riding for Deliveroo: Resistance in the new economy.*
Polity. For an example of slow-down in yet another industry (academia)
see Vostal, F., Benda, L., and Virtová, T. (2019). Against reductionism: On
the complexity of scientific temporality. *Time & Society, 28*(2), 783–803.
10. Parker, C. (2017, November 27). Amazon warehouse life "revealed with
timed toilet breaks and workers sleeping on their feet." *The Sun.*
11. Allen, S. (2018, July 31). *Some thoughts* [Video]. *YouTube.*
12. Bruder, J. (2019, November 12). Meet the immigrants who took on
Amazon. *Wired.*
13. Laguerre, M. (2003). The Muslim chronopolis and diasporic temporality.
Research in Urban Sociology, 7, 57–81.
14. As described by a former Amazon executive. See Kantor, J., Weise, K., and
Ashford, G. (2021, June 15). The Amazon that customers don't see. *The
New York Times.*
15. Slade, G. (2007). *Made to break.* Harvard University Press, p. 5.

16. Elcioglu, E.F. (2010). Producing precarity: The temporary staffing agency in the labor market. *Qualitative Sociology, 33*(2), 117–136.

17. Tung, I., and Berkowitz, D. (2020, March 6). *Amazon's disposable workers: High injury and turnover rates at fulfillment centers in California.* National Employment Law Project data brief, p. 2.

18. Romano, B. (2020, October 10). Amazon's turnover rate amid pandemic is at least double the average for retail and warehousing industries. *The Seattle Times.*

19. Soper, S. (2020, October 6). Amazon study of workers' COVID is faulted over lack of key data. *Bloomberg.*

20. Mojtehedzadeh, S. (2021, March 21). More than 600 Amazon workers in Brampton got COVID-19. Why were so few reported to the province? *The Toronto Star.*

21. As discussed in Chapter 2.

22. For an example from 1960s Italian factories see the pages on Olivetti in Alquati, R. (1975). *Sulla FIAT e altri scritti.* Feltrinelli.

23. A discussion of the phenomenon at Amazon can be found in Semuels, A. (2018, February 14). Why Amazon pays some of its workers to quit. *The Atlantic.*

24. Bezos, J. (2019). *2018 letter to shareholders.* Retrieved from www.aboutamazon.com.

25. Bezos, J. (2014). *2013 letter to shareholders.* Retrieved from www.aboutamazon.com.

26. Taylor, B. (2008, May 19). Why Zappos pays new employees to quit—and you should too. *Harvard Business Review.*

27. Amazon (2021) What is Amazon's Career Choice? Retrieved from www.aboutamazon.com.

28. Alquati, R. (1975). *Sulla FIAT e altri scritti.* Feltrinelli, p. 146.

29. Doeringer, P.B., and Piore, M.J. (1985). *Internal labor markets and manpower analysis.* Routledge.

30. Marx, K. (1867, 1976). *Capital* (Vol. 1). Penguin, pp. 546–547.

31. Lepak, D.P. and Snell, S.A. (1999). The human resource architecture: Toward a theory of human capital allocation and development. *Academy of Management Review, 24*, 31–48.

32. Wright, M.W. (2006). *Disposable women and other myths of global capitalism.* Taylor & Francis, p. 2.

33. Wright, M.W. (2006). *Disposable women and other myths of global capitalism.* Taylor & Francis, pp. 25–26.

34. Sterne, J. (2007). Out with the trash: On the future of new media. In: Acland, C. (ed.), *Residual media.* University of Minnesota Press, p. 23.

35. On workers being actively "produced" see also Salzinger, L. (2003). *Genders in productions: Making workers in Mexico's global factories.* University of California Press.

36. Appadurai, A., and Neta, A. (2020). *Failure.* Polity, p. 2.

37. So workers must become flexible, fit and legible by surveillance, as discussed in Chapter 3. They must develop the "perverse virtues of digital labor," as put by Gregory, K., and Sadowski, J. (2021). Biopolitical platforms: The perverse virtues of digital labour. *Journal of Cultural Economy*. On the pamphlet itself see Ongweso, E. (2021, June 1). Amazon calls warehouse workers "industrial athletes" in leaked wellness pamphlet. *VICE*.

38. Sharma, S. (2014). *In the meantime: Temporality and cultural politics.* Duke University Press.

39. Cowen, D. (2014). *The deadly life of logistics: Mapping violence in global trade.* University of Minnesota Press, p. 113.

40. For a review of this phenomenon see Gerstel, N., and Clawson, D. (2018). Control over time: Employers, workers, and families shaping work schedules. *Annual Review of Sociology, 44*, 77–97.

41. Appadurai, A., and Neta, A. (2020). *Failure.* Polity.

5 REIMAGINE NOW

1. Montfort, N. (2017). *The future*. MIT Press, p. 4.

2. On technology and other fixes see Harvey, D. (2005). *A brief history of neoliberalism*. Oxford University Press.

3. Bezos, J. (2021). *2020 letter to shareholders*. Retrieved from www.aboutamazon.com.

4. Nicas, J. (2018, March 22). At Mars, Jeff Bezos hosted roboticists, astronauts, other brainiacs and me. *The New York Times*.

5. Berg, P., Isaacs, P.W., and Blodgett, K. (2016). Airborne fulfillment center utilizing unmanned vehicles for item delivery. US Patent No. 9,305,280.

6. See www.amazonrobotics.com.

7. On automation and technological unemployment see Wajcman, J. (2017). Automation: Is it really different this time? *The British Journal of Sociology, 68*(1), 119–127. On Amazon workers' reactions see Reese, E., and Struna, J. (2018). "Work Hard, Make History": Oppression and resistance in Inland Southern California's warehouse and distribution industry. In: Alimahomed-Wilson, J., and Ness, I. (eds.), *Choke points. Logistics workers disrupting the global supply chain*. Pluto Press, pp. 81–95.

8. Selin, C. (2008). The sociology of the future: Tracing stories of technology and time. *Sociology Compass, 2*(6), 1885.

9. Johns, A. (2009). *Piracy: The intellectual property wars from Gutenberg to Gates*. The University of Chicago Press, p. 426.

10. For a thorough analysis of such spheres see Coombe, R. (1998). *The cultural life of intellectual properties*. Duke University Press.

11. Burk, D.L., and Reyman, J. (2014). Patents as genre: A prospectus. *Law & Literature, 26*(2), 163–190.

12. Hetherington, K. (2017). Surveying the future perfect: Anthropology, development and the promise of infrastructure. In: Harvey, P., Morita, A., and Jensen, C.B. (eds.), *Infrastructures and social complexity: A companion*. Routledge, pp. 40–50.

13. Urry, J. (2016). *What is the future?* Polity, p. 189.

14. On R&D investments see data compiled by Szmigiera, M. (2021, March 17). Ranking of the 20 companies with the highest spending on research and development in 2018. *Statista*.

15. Rikap, C. (2020). Amazon: A story of accumulation through intellectual rentiership and predation. *Competition & Change, 0*(0), 1–31.

16. Hartman, P., Bezos, J., Kaphan, S., and Spiegel, J. (1999). Method and system for placing a purchase order via a communications network, US Patent No. 5,960,411. See also Stone, B. (2013). *The everything store: Jeff Bezos and the age of Amazon*. Little, Brown and Company, pp. 76–77.

17. Thomson Reuters (2019, May 13). Amazon to introduce more automated packaging machines. *CBC*.

18. SpekWork. (2018). *GigCo*. Retrieved from http://spek.work/.

19. For a history of the Luddite movement and its relevance for today's technology see Sale, K. (1996). *Rebels against the future: The Luddites and their war on the industrial revolution*. Basic Books.

20. Vonnegut, K. (1952). *Player piano*. Scribner.

21. For instance, Theodore Roszak discussed the portrayal of "technocratic despotism" in the novel, while David Noble said that the technology described by Vonnegut was actually meant to augment rather than replace workers' skills. See Roszak, T. (1994). *The cult of information: A neo-Luddite treatise on high-tech, artificial intelligence, and the true art of thinking*. University of California Press; and Noble, D. (1984). *Forces of production: A social history of industrial automation*. Oxford University Press.

22. Tubaro, P., and Casilli, A.A. (2019). Micro-work, artificial intelligence and the automotive industry. *Journal of Industrial and Business Economics, 46*(3), 333–345.

23. Stallman, T., Brady, T.M., Bocamazo, M.R., Borges, M.G., Davidson, H.S., Johnson, A.R., Rodrigues, A.P., and Tieu, M. (2019). Modular automated inventory sorting and retrieving. US Patent No. 10,217,074.

24. Crawford, K., and Joler, V. (2018). *Anatomy of an AI System*. Retrieved from www.anatomyof.ai.

25. See Danaher, J. (2015). The threat of algocracy: Reality, resistance and accommodation. *Philosophy of Technology, 29*, 245–268; and LeCavalier, J. (2016). *The rule of logistics: Walmart and the architecture of fulfillment*. University of Minnesota Press.

26. Koka, M., Raghavan, S.N., Asmi, Y.B., Chinoy, A., Smith, K.J., and Kumar, D. (2019). Using gestures and expressions to assist users. US Patent No. 10,176,513.

27. Curlander, J.C., Graybill, J.C., Madan, U., Tappen, M.F., Bundy, M.E., and Glick, D.D. (2018). Selecting items for placement into available volumes using imaging data. US Patent No. 9,864,911.

28. McNamara, A.M., Smith, K.J., Hollis, B.R., Boyapati, S., and Frank, J.J. (2019). Color adaptable inventory management user interface. US Patent No. 10,282,695.

29. Bezos, J. (2006). *2005 letter to shareholders*. Retrieved from www.aboutamazon.com.

30. See Lee, M.K., Kusbit, D., Metsky, E., and Dabbish, L. (2015). Working with machines: The impact of algorithmic and data-driven management on human workers. *Proceedings of the 33rd Annual ACM Conference on Human Factors in Computing Systems*, USA, pp. 1603–1612, Chen, J. (2017). Thrown under the bus and outrunning it! The logic of didi and taxi drivers' labour and activism in the on-demand economy. *New Media & Society, 20*(8), 2691–2711, and Wood, A., Graham, M., Lehdonvirta, V., and Hjorth, I. (2019). Good gig, bad gig: Autonomy and algorithmic control in the global gig economy. *Work, Employment & Society, 33*(1): 56–75.

31. Aneesh, A. (2009). Global labor: Algocratic modes of organization. *Sociological Theory, 27*(4), 347–370; Danaher, J. (2015). The threat of algocracy: Reality, resistance and accommodation. *Philosophy of Technology, 29,* 245–268.

32. Mountz, M.C., Glazkov, O., Bragg, T.A., Verminski, M.D., Brazeau, J.D., Wurman, P.R., Cullen, J.W., and Barbehenn, M.T. (2015). Inter-facility transport in inventory management and fulfillment systems. US Patent No. 8,972,045.

33. Madan, U., Bundy, M.E., Glick, D.G., and Darrow, J.E. (2018). Augmented reality user interface facilitating fulfillment. US Patent No. 10,055,645.

34. Panzieri, R. (1961). Sull'uso capitalistico delle macchine nel neocapitalismo. *Quaderni rossi, 1,* pp. 53–72.

35. Lopez, G.E., Walsh, P.J., McMahon, J.A., and Ricci, C.M. (2015). Fulfillment of orders from multiple sources. US Patent No. 9,195,959.

36. Madan, U., Bundy, M.E., Glick, D.G., and Darrow, J.E. (2018). Augmented reality user interface facilitating fulfillment. US Patent No. 10,055,645.

37. Bettis, D., McNamara, A., Hollis, B., Étienne, F., Boyapati, P., Smith, K.J., and Jones, J.B. (2019). Augmented reality enhanced interaction system. US Patent No. 10,282,696.

38. Panzieri, R. (1967). Lotte operaie nello sviluppo capitalistico. *Quaderni piacentini, 6*(29), 37.

39. Yarlagadda, P.K., Archambeau, C., Curlander, J.C., Donoser, M., Herbrich, R., O'Brien, B.J., and Tappen, M.F. (2018). System for configuring a robotic device for moving items. US Patent No. 10,099,381.

40. Brady, T.M. (2018). Wrist band haptic feedback system. US Patent No. 9,881,277 B2.

41. Brazeau, J.D., and Mendola, S. (2017). Inventory event detection using residual heat information. US Patent No. 9,767,432.

42. Wellman, P.S., Verminski, M.D., Stubbs, A., Shydo Jr., R.M., Claretti, E., Aronchik, B., and Longtine, J.G. (2017). Robotic grasping of items in the inventory system. US Patent No. 9,561,587.

43. Gupta, B., Aalund, M.P., and Mirchandani, J. (2019). Method and system for tele-operated inventory management system. US Patent No. 10,464,212 B2.

44. See Chapter 4 for other examples of such "virtual migrations."

45. Rouaix, F., Antony, F.F., Elliott, C.L., and Bezos, J.P. (2017). Light emission guidance. US Patent No. 9,852,394.

46. Wurman, P., Brazeau, J.D., Farwaha, P.S., Holt, R.A., Durham, J.W., Enright, J.J., Glazkov, A., and Holcomb, J.B. (2016). Re-arrange stored inventory holders. US Patent No. 9,452,883.

47. Durham, J.M., Dresser, S., Longtine, J.G., Mills, D.G., Wellman, P.S., and Wilson, S.A. (2019). Amassing pick and/or storage task density for inter-floor transfer. US Patent No. 1,0395,152.

48. On the use of AI for emotional recognition purposes see Crawford, K. (2021). Time to regulate AI that interprets human emotions. *Nature*, *592*(7853), 167.

49. Newton, C. (2017, June 12). A boredom detector and 6 other wild Facebook patents. *The Verge*.

50. Koka, M., Raghavan, S.N., Asmi, Y.B., Chinoy, A., Smith, K.J., and Kumar, D. (2019). Using gestures and expressions to assist users. US Patent No. 10,176,513.

51. Atanasoski, N., and Vora, K. (2019). *Surrogate humanity: Race, robots, and the politics of technological futures*. Duke University Press; for an analysis of recent automation in the domestic sphere see Fortunati, L. (2018). Robotization and the domestic sphere. *New Media & Society*, *20*(8), 2673–2690.

52. WIRED UK. (2017, May 16). Inside Ocado's distribution warehouse [Video]. *YouTube*.

53. Spiegel, J., McKenna, M., Lakshman, G., and Nordstrom, P. (2013). Method and system for anticipatory package shipping. US Patent 8,615,473 B2.

54. Amoore, L. (2011) Data derivatives: On the emergence of a security risk calculus for our times. *Theory, Culture & Society 28*(6), 24–43. On the predictive nature of algorithms see also Arvidsson, A. (2016) Facebook and finance: On the social logic of the derivative. *Theory, Culture & Society 33*(6), 3–23; and Gillespie, T. (2014) The relevance of algorithms. In: Boczkowski, P., Foot, K., and Gillespie, T. (eds.). *Media technologies: Essays on communication, materiality, and society*. MIT Press, pp. 167–194.

55. LeCavalier, J. (2016). *The rule of logistics: Walmart and the architecture of fulfillment*. University of Minnesota Press.

56. Licklider, J.C.R. (1960). Man-computer symbiosis. *IRE Transactions on Human Factors in Electronics, 1,* 4.

57. LeCavalier, J. (2016). *The rule of logistics: Walmart and the architecture of fulfillment.* University of Minnesota Press, p. 152; on this point see also Autor, D. (2015). Why are there still so many jobs? The history and future of workplace automation. *Journal of Economic Perspectives, 29*(3), 3–30.

58. Mosco, V. (2005). *The digital sublime: Myth, power, and cyberspace.* MIT Press.

59. Wajcman, J. (2017). Automation: Is it really different this time? *The British Journal of Sociology, 68*(1), 119–127.

60. Stubbs, A., Verminski, M.D., Caldara, S., and Shydo Jr., R.M. (2018). System and methods to facilitate human/robot interaction. US Patent No. 9,889,563.

61. Some have theorized that a new kind of "inhuman power" stemming from the widespread application of artificial intelligence to our societies would bring about major qualitative change in the relation between workers and capital. See Dyer-Witheford, N., Kjøsen, A.M., and Steinhoff, J. (2019). *Inhuman power: Artificial intelligence and the future of capitalism.* Pluto Press, p. 58.

62. Urry, J. (2016). *What is the future?* Polity Press, p. 12.

63. Atanasoski, N., and Vora, K. (2019). *Surrogate humanity: Race, robots, and the politics of technological futures.* Duke University Press, p. 13.

64. Chassany, A. (2016, March 3). Uber: A route out of the French banlieues. *Financial Times.*

65. Noble, S. (2018). *Algorithms of oppression: How search engines reinforce racism.* NYU Press.

6 MAKE HISTORY

1. Panzieri, R. (1976). *Lotte operaie nello sviluppo capitalistico.* Einaudi.

2. On decomposition, especially in relation to the use of technology, as well as on struggles in the digital economy see Dyer-Witheford, N., Kjøsen, A.M., and Steinhoff, J. (2019). *Inhuman power: Artificial intelligence and the future of capitalism.* Pluto Press.

3. On Amazon's environmental politics see, for instance, Caraway, B. (2020). Interrogating Amazon's sustainability innovation. In: Oakley, K., and Banks, M. (eds.). *Cultural industries and the environmental crisis.* Springer, pp. 65–78.

4. See Mak, A. (2021, March 2). Amazon's anti-union campaign is going to some strange places. *Slate.* The website #DoItWithoutDues is still visible at https://web.archive.org/web/20210329170825/https://www.doit withoutdues.com/.

5. For a comparative analysis of labor union politics across different countries see a number of chapters in Alimahomed-Wilson, J., and Reese, E.

(2020). *The cost of free shipping: Amazon in the global economy.* Pluto Press; and several pieces in Transnational Social Strike (2019). *Strike the giant! Transnational organization against Amazon.* Retrieved from www. transnational-strike.info.

6. For instance, the Amazon Alliance and Amazon Workers International (see www.uniglobalunion.org and https://amworkers.wordpress.com).

7. CGIL and CISL are the biggest union federations in Italy. CGIL is the former communist union, while CISL was historically linked to the Christian Democratic party.

8. As suggested among others by Gent, C. (2020, November 27). How do we solve a problem like Amazon? *Novara Media.*

9. For instance see Amazon Workers Collective (2020, October 28). Workers of the world: Salt at Amazon! *Briarpatch.*

10. See Brecher, J. (2014). *Strike!* PM Press.

11. For experiences in this direction see Transnational Social Strike (2019). *Strike the Giant! Transnational organization against Amazon.* Retrieved from www.transnational-strike.info.

12. Day, M., Lepido, D., Fouquet, H., and Munoz Montijano, M. (2020, March 16). Coronavirus strikes at Amazon's operational heart: Its delivery machine. *Bloomberg.*

13. Scott, J. (1987). *Weapons of the weak: Everyday forms of peasant resistance.* Yale University Press.

14. The quote is from the blog post he used to announce his decision: Bray, T. (2020). *Bye, Amazon.* Retrieved from www.tbray.org.

15. Karppi, T., and Nieborg, D.B. (2020). Facebook confessions: Corporate abdication and Silicon Valley dystopianism. *New Media & Society,* 1461444820933549.

16. Foucault M. (1978). *The history of sexuality volume I.* Allen Lane.

17. See Smith, B. (2020). Thanks Amazon for scarring me for life. Worker breakdown and the disruption of care at Amazon. *AoIR Selected Papers of Internet Research.* Retrieved from www.aoir.org.

18. As described in Adler-Bell, S. (2019). Surviving Amazon. *Logic* 8. Retrieved from www.logicmag.io. For concrete examples see www.reddit. com/r/AmazonFC.

19. See www.axel-springer-award.com.

20. Paul, K. (2019, May 23). Amazon workers demand Bezos act on climate crisis. *The Guardian.*

21. In Italian law, a state of agitation can be formally declared and provides extra protection to worker action during negotiations.

22. Alimahomed-Wilson, J., and Ness, I. (eds.) (2018). *Choke points. Logistics workers disrupting the global supply chain.* Pluto Press.

23. As described among others by Bernes, J. (2014). *Logistics, counterlogistics and the communist prospect.* Retrieved from www.endnotes.org.uk. See also material produced by the Into the Black Box research collective at www.intotheblackbox.com.

24. Including in Italy. See Gruppo Nord Est Di Inchieste Dal Basso (2021, February 4). L'offensiva di Amazon nel Nord Est. *GlobalProject*.

25. Most importantly the coalition Athena, see www.athenaforall.org.

26. On emerging forms of resistance within digital capitalism see for instance Huws, U. (2014). *Labor in the global digital economy: The cybertariat comes of age*. Monthly Review Press; or for a broad survey of cases see Delfanti, A., and Sharma, S. (2020). Log out! The platform economy and worker resistance. *Notes from Below*, 8.

27. On such conditions of possibility, see Van Doorn, L. (2019). On the conditions of possibility for worker organizing in platform-based gig economies. *Notes from Below*, 8.

28. For instance, see Cant, C., and Mogno, C. (2020). Platform workers of the world, unite! The emergence of the transnational federation of couriers. *South Atlantic Quarterly*, *119*(2), 401–411; or Qadri, R. (2020). Algorithmized but not atomized? How digital platforms engender new forms of worker solidarity in Jakarta. *Proceedings of the AAAI/ACM Conference on AI, Ethics, and Society*, p. 144.

29. Liu, W. (2020). *Abolish Silicon Valley: How to liberate technology from capitalism*. Repeater.

30. Phillips, L., and Rozworski, M. (2019). *People's Republic of Walmart. How the world's biggest corporations are laying the foundation for socialism*. Verso.

Index

The Pluto Press Newsletter

Hello friend of Pluto!

Want to stay on top of the best radical books
we publish?

Then sign up to be the first to hear about our
new books, as well as special events,
podcasts and videos.

You'll also get 50% off your first order with us
when you sign up.

Come and join us!

Go to bit.ly/PlutoNewsletter